"David Robertson has done it again. *The Power of Little Ideas* draws on rich examples—from car sales to toy making to digital technology—to provide a powerful, practical approach to innovation resulting in sustainable success for the company."

—**STEPHEN K. KLASKO, MD, MBA,** President and CEO, Thomas Jefferson University and Jefferson Health; coauthor, *We CAN Fix Healthcare*

"David Robertson demonstrates a compelling new innovation space between incremental and radical innovations. This 'Third Way' is an excellent new resource for innovation practitioners in any industry."

—**MICK SIMONELLI,** former Lead Innovation Executive, USAA

"This important book is packed with useful case studies, thoughtful advice, and a refreshingly clear-eyed view of how different forms of innovation work today. It's perfect for business leaders and managers who want to innovate in a more considered and realistic way for their business and brand."

—**TOM ANDREWS,** President, SYPartners Consulting

THE POWER OF LITTLE IDEAS

David Robertson
with Kent Lineback

THE POWER OF LITTLE IDEAS

A Third Way to Innovate for Market Success

Harvard Business Review Press
Boston, Massachusetts

Printed in the United States of America

10 9 8 7 6 5 4 3 2

Library of Congress Cataloging-in-Publication Data
Names: Robertson, David C. (David Chandler), author. | Lineback, Kent, 1943- author.
Title: The power of little ideas : a low-risk, high-reward approach to innovation / by David C. Robertson with Kent Lineback.
Description: Boston, Massachusetts : Harvard Business Review Press, [2017]
Identifiers: LCCN 2016049137 | ISBN 9781633691681 (hardcover : alk. paper)
Subjects: LCSH: Technological innovations. | New products. | Management.
Classification: LCC HD45 .R595 2017 | DDC 658.4/063—dc23 LC record available at https://lccn.loc.gov/2016049137

ISBN: 9781633691681
eISBN: 9781633691698

The paper used in this publication meets the requirements of the American National Standard for Permanence of Paper for Publications and Documents in Libraries and Archives Z39.48-1992.

CONTENTS

THE POWER OF LITTLE IDEAS

Preface

Seeing Innovation Differently

I wrote this book, at least in part, because of what I learned a few years ago from Stephen, the guy who painted my house. We hired Stephen and his crew because he had done great work on many neighbors' houses. We weren't disappointed—he did an excellent job on mine, and I would recommend him to anyone. He not only painted the house but also helped my wife and me decide which colors to use, fixed the gutters, and did many other small repairs around the house. But hiring Stephen and his crew also illustrated something about innovation and gave me yet another reason to write this book.

When Stephen submitted his estimate to me for the work, he told me he planned to use Sherwin-Williams paint. He explained that it was a good, high-quality paint, but also that he preferred Sherwin-Williams because of the service the company provided him. He told me how easy it was to work with Sherwin-Williams, how close the local Sherwin-Williams store was, how the salespeople would help us with color selections, and that the store gave him free same-day delivery if he ran out of paint or supplies.

I did a quick map search of Sherwin-Williams and was surprised by what I found. Within a five-minute drive of my house in suburban Philadelphia, there were five Sherwin-Williams paint stores (six, if you count one dedicated to automotive finishes). There are only three Starbucks cafés within the same distance.

I stopped by the Sherwin-Williams store in my neighborhood and met Tom, the Sherwin-Williams sales rep who works with Stephen, to understand why Stephen liked the company. Beyond the many locations and good-quality paint, Tom and others from Sherwin-Williams also support Stephen throughout the process. Tom will come out to the job site to help Stephen estimate the job. He'll help Stephen develop an accurate plan for each phase of the project, and he'll help Stephen estimate how much labor and material will be needed at each stage. The rep will also check Stephen's proposal to make sure that all the necessary materials and equipment are included, then make sure that the right amount of material is available when it is needed. Tom starts his day an hour before his store opens, because that's when his customers—the painting contractors—start their days.

During the job, Tom allows Stephen to adjust the amount of paint as needed. For example, if Stephen buys 50 percent too much primer, he can return the unused cans for a full credit, and Sherwin-Williams will adjust the amount of finish-coat paint for the job, preventing waste and extra expense (while primer paint can usually be returned, custom colored paint for the finish coat can't). Tom will check the daily orders and suggest items that Stephen might have forgotten. At the end of the job, Tom will help Stephen write up an estimate for the next job (and, as any homeowner knows, there's always a next job).

As I looked around the Sherwin-Williams store, I saw the expected paint color displays and marketing brochures. Near the entrance was an offer for a color consultant to come to my house to help me choose paint colors. On the wall was a full range of brushes, tools, and gadgets to help painters. While I was waiting for Tom, the sales clerk showed me a device that makes it easy for painters to cover the bottom of their shoes when they come in the door, and she told me about a tool that automatically applied mud to drywall tape, reducing materials use and labor cost. Tom told me about the annual show he attends to learn about new devices, techniques, and products for painters. He explained the contractor program that rewards painters like Stephen with increasing discounts as the volume of business increases.

Sherwin-Williams is proud of the quality of its paint. But my house isn't covered with Sherwin-Williams paint because I thought the paint was better. In fact, the consumer ratings magazine I trust recommended a slightly higher-rated paint at half the price. But Stephen explained that only about 15 percent of the cost of a painting job was the paint itself, and if I wanted him to use a different paint, the cost of the project would be higher.

Sherwin-Williams isn't selling paint. It's selling a complete service to small painting companies like Stephen's. It supports those small businesses with a complete end-to-end service, and it's doing quite well in the process: sales in 2015 are up 46 percent over 2010, and profits are up 128 percent in the same period. The company is constantly innovating in its core product—including a microbicidal paint that kills common bacteria like *Staphylococcus* and *E. coli* on contact. But it's also innovating in the services it provides to painting contractors and the complementary products that it sells in the stores.

This approach to innovation doesn't fall neatly into the usual categories that we see in the business press. Sherwin-Williams isn't disrupting the paint industry; nor is it sailing for blue oceans or acting like a lean startup. The company is not revolutionizing the future of house painting or simply improving its core paint products. Its approach to innovation is unique.

In my previous book, *Brick by Brick: How LEGO Rewrote the Rules of Innovation and Conquered the Global Toy Industry*, I told the story of how LEGO adopted a similar innovation approach in 2003 to recover from its brush with bankruptcy. Like Sherwin-Williams's success, LEGO's recovery and growth haven't come from just offering a better core product or from reinventing the future of its industry. In fact, LEGO tried both of those strategies and failed. The successful strategy for the toymaker was to go back to the company's core, the box of bricks, understand what the customer wanted from that product, and innovate around the box. This approach to innovation, neither incremental improvement in current products nor revolutionary disruption of those products, is something we'll call the *Third Way* to innovate, and it's not being discussed or codified in the literature. The goal of this book is to define and explain this approach.

When LEGO mastered this approach, the company recovered quickly and spectacularly. When LEGO posted its annual results in early 2016, their eight-year average annual sales growth was 21 percent per year, and profit growth an equally impressive 36 percent per year. Given that the patents for the brick expired in the 1980s and aggressive competitors make LEGO-compatible bricks for a fraction of the price, this growth is nothing short of astounding.

After immersing myself in the world of LEGO and writing the LEGO book, I began to see this same approach in many other places. Not only is the approach working for Sherwin-Williams, but it has also spurred growth at CarMax, Gatorade, USAA, and Victoria's Secret. And the Apple Computer turnaround story, which began with Steve Jobs's return to Apple in 1997, followed a process very similar to LEGO's. These companies' strategies seemed to follow a pattern—a set of steps they took as they tried this type of innovation.

The goal of this book is to show you, the reader, how you can learn and adopt the approach that LEGO and others have used so successfully, without the crisis that precipitated LEGO's turnaround. The other companies that have adopted the LEGO strategy followed a sequence of decisions, a process that is the focus of this book. This book lays out the steps you should follow and the challenges you'll face if you decide to adopt this approach.

As host of *Innovation Navigation*, a weekly radio show and podcast (www.innonavi.com), I try to read every innovation book that is published. Surprisingly, relatively few of these books are relevant for people whose job is to innovate in an existing market. Most people in most companies are focused on making existing products more attractive to existing customers, but there are remarkably few innovation books focused on this type of innovation. Too often, innovation gurus tell people to take a clean-sheet approach to innovation, to start from scratch and create something insanely great. While this may be good advice for a company, it's rarely helpful for a product manager or a business unit head whose jobs require them to make a current product better.

This book is meant for anyone whose job is to extract maximum value from an important product. People responsible for delivering an existing product to existing customers have a difficult job—they're tremendously constrained and they're often under a great deal of pressure. A central goal of this book is to define a unique and powerful approach to innovation and to help innovators navigate the accompanying challenges. If you're in this position—if your job is to keep a current product fresh and relevant in a competitive market—this book is for you.

But this book will also be useful for those developing new types of products or taking existing products into new markets. Exploring such new frontiers is a tremendous challenge, and we hope that these innovators will also find useful ideas about how to make new products successful.

1

How Little Innovations Produce Big Results

When Sarah Robb O'Hagan, a general manager at Nike, agreed to assume leadership of Gatorade in 2008, she thought she was taking over an iconic brand that had grown a little tired. But when she arrived at Gatorade headquarters in Chicago in July of that year, what she found was something else—a struggling brand in obvious decline.

Gatorade had invented the sports drink category in the 1960s. But in 2007, sales had stalled, and in the fifty-two weeks preceding Robb O'Hagan's arrival, they had actually dropped 10 percent, while sales of cheaper archrival Powerade had grown 13 percent. The Gatorade product and marketing team she took over was already rushing to redesign Gatorade's logo and packaging.

What would you do if you were in her position? Clearly, innovation is the key. But how can you innovate to revive the brand and restart growth?

One common response is more—more products for more customers, more features or more performance for current products, and more channels of distribution or expansion into more geographic markets. This is usually the first and easiest strategy to revive sales. Unfortunately for Robb O'Hagan, Gatorade had already tried this approach. Seven years before, in 2001, when PepsiCo bought it, Gatorade expanded its range of flavors, added low-calorie versions, and put the drink through Pepsi's massive distribution network. Sales took off.

But "more" has limits. Expanding distribution and adding new product variants will quickly hit a point of diminishing returns, after which each addition generates fewer marginal sales but just as much additional cost. When taken too far, new product versions begin to lose money, and, eventually, there are no new distribution channels to fill. For Gatorade, 2007 was the year it hit those limits.

What were the Gatorade team's choices? Sustaining and incremental innovations—new versions, new channels, and the like—were exhausted.*

Conventional thinking about innovation would point the Gatorade team in a different direction. It says the only real alternative to sustaining and incremental innovation is to go big. Hundreds of articles and books, with more appearing every year, explain how to pursue revolutionary, radical change: look for "blue oceans," develop new "disruptive" products or business models, turn your product into a memorable experience, or act like a lean startup.

Of course, there are differences among these approaches, as their advocates will quickly point out. In spite of those differences, however,

*A *sustaining innovation* is one intended to sustain or preserve a product's sales and market share. Thus, *sustaining* refers to the purpose of the innovation. In theory, sustaining innovations can be large or small, though typically they're small enhancements to a product. An *incremental innovation* is a small change or improvement in a product. *Incremental* refers to the scale or magnitude of the innovation. An innovation can be, and often is, both sustaining and incremental, but the two terms are not strictly equivalent.

they all share certain basic features. All promise dramatic growth and all agree: more of the same won't cut it. They all tell us to look at the icons of radical innovation transforming the world—digital photography replacing film, Uber replacing taxis, online news supplanting newspapers, Airbnb replacing hotels, and Amazon replacing everyone else.

For our purposes, though, the key feature they all share is this: when pursued by existing organizations, these approaches often lead firms to rethink their businesses in fundamental ways. They typically call for large and risky investments, not just in money but also in time, effort, and strategic focus. And because these approaches typically take organizations into new territory—new technology, new products, new markets, and new processes—the full consequences they produce are often unforeseeable, failure is common, and the cost of failure is large. One careful evaluation of revolutionary innovations estimates that failure rates are 60–75 percent, as opposed to 25–40 percent for incremental improvements.[1]

Because of these similarities, we label all these forms of innovation *radical*. Whatever external form they may take, they are all, for incumbents, internally disruptive. All of them rest on the same underlying assumption: to succeed in today's hypercompetitive global economy, you must respond to competitive threats by changing your business in some fundamental way. If you don't, someone else will disrupt it for you.*

The consequence of this binary thinking—if incremental doesn't work, do something radical—is that many companies respond to the first sign of a threat by saying, "We have to do something new, big and revolutionary." So they launch a major initiative and challenge

*Christensen and his colleagues have a very specific definition of disruption that is more narrow than the meaning used in this book. For the purposes of this book we will use *disruptive* in its more commonly accepted definition as an innovation that is new to its industry and has a significant impact on its market.

their employees (or expensive consultants) to think far outside the box. A flurry of big new ideas emerges, followed by new initiatives and Skunk Works teams. But the success rate is low. The entrenched processes, systems, training, and values that produced and sustained prior success for those companies now conspire to make them less successful at radical change. If they're fortunate, after the dust settles, they've only wasted time and money. Worst case, they've put in place drastic changes with unintended consequences that cannot be entirely undone. In either case, they've poisoned the well for more big new ideas.

Where does this leave Sarah Robb O'Hagan and her team at Gatorade in 2008? If more and better versions of Gatorade, along with increased distribution, weren't enough, was radical innovation their only alternative? Was that their only path to restarting growth?

No, it wasn't. They chose a different path that produced one of the great brand turnarounds of the new century. Their story is one of several we will tell about leaders and companies that have refused to accept today's binary thinking. For some time now, it has been clear that neither sustaining nor radical innovation can explain a number of corporate success stories—including some, like Apple after Steve Jobs returned in 1997, that have mistakenly been held up as exemplars of disruptive innovation. The experiences of companies such as Gatorade, Apple, LEGO, Victoria's Secret, Guinness, Novo Nordisk, and CarMax—all stories that we will tell—reveal a third option that's often less risky and less costly than radical change but is in many cases equally powerful.

It is this third option that Robb O'Hagan and her team used to revive Gatorade. For convenience, we will refer to this option as the Third Way simply to indicate it's not bound by the binary thinking that says innovators have only two choices: innovate small or innovate big. There is another option.

The Turnaround at Gatorade

Gatorade was no ordinary soft drink. Scientists at the University of Florida had developed it in the 1960s as a hydration aid for the school's football players who had to play under the brutal Florida sun. It quickly became a favorite of athletes everywhere, and in 1983, it was named the official sports drink of the National Football League.

PepsiCo bought Gatorade in 2001, introduced a raft of new flavors and other variations, and pushed it through the vast Pepsi distribution system as a soft drink for the mass beverage consumer. That approach hyped growth for a few years; by 2007, Gatorade commanded 80 percent of the $8 billion sports drink market in the United States. But when the economy began to falter, sales growth disappeared. And when the Great Recession arrived in 2008, three months after Robb O'Hagan joined Gatorade, sales went south in a hurry.

As a first step, Robb O'Hagan helped her team finish its redesign of the logo and packaging. In the redesign, which appeared in early 2009, *Gatorade* became simply *G* with a more up-to-date lightning-strike design. The introduction of the new design didn't go well, however. On retailers' shelves, old and new designs were mixed side by side, and sales continued to shrink. Retailers and Wall Street analysts alike blamed the new design and were quick to note that TV ads preceding the 2009 Super Bowl featured the new design but never mentioned the name "Gatorade" (though the ad that ran during the game did include it). Concerned voices inside PepsiCo called for returning to the old design and doubling down the old strategy of head-to-head competition with Powerade, the way Pepsi had always competed with Coke.

Instead of panicking, the Gatorade team looked at the market data streaming in. It told them that Gatorade was losing casual drinkers, many if not most of them the customers added since PepsiCo had acquired the brand. These casual drinkers were going elsewhere, many to plain old tap water, not a bad move when the economy was tanking for those who never had any real reason to drink Gatorade in the first place. The good news in the data was that serious athletes were sticking with Gatorade.

Gatorade's core customers, those serious athletes, came in two basic flavors: teenage athletes keen to win and older athletes such as marathoners and triathletes. The teenagers accounted for 15 percent of Gatorade's customers, and the older athletes another 7 percent; together, that 22 percent accounted for 46 percent of Gatorade sales.

In recent years, Gatorade hadn't been marketing as deeply to those loyal segments. It was being distributed through mass-market outlets like convenience stores, grocery stores, and big discounters, but not through runners' stores, cycling shops, and other specialized retailers that serious customers frequented.

What it all meant was that Gatorade faced a choice: compete with Powerade on price, which didn't make much sense, or refocus on serious teen and older athletes and mostly ignore everyone else. Refocusing, however, would take more than a redesigned logo and a catchy advertising campaign.

As Robb O'Hagan and her group looked hard at these core customer segments, they realized that truly serving them meant going beyond hydration. "Athletes needed a full range of specialized nutrition," she said, "and they needed it before, during, and after the event. We called it 'Sports Fuel' and discovered that even world-class athletes didn't know where to find it. Usain Bolt, the great runner, ate Skittles candies before Olympic races . . . because he couldn't find anything better designed for his needs."[2]

Their goal was to make Gatorade the "aspirational brand for athletes" it had once been. Their plan comprised two basic steps:

First, refocus Gatorade on serious athletes (athletic adolescents and older performance athletes) who were Gatorade's sweet spot.

Second, expand Gatorade products to supply all the fuel—hydration and nutrition—that serious athletes needed. The *G* rebranding gave Robb O'Hagan and her team the flexibility to move beyond fluids. In spite of problems during the switchover, research told them that younger athletes in particular understood and accepted the change.

Their plan reversed the typical approach of mass beverage marketing. Instead of taking the same basic product to more and more different customers, they added a new line of products and aimed them all, including the Gatorade drink, at a specific customer segment, serious athletes. And they were going to do it in spite of pressure inside and outside PepsiCo to return to tried-and-true mass-marketing approaches. In taking this new direction, the team rationalized the line of drinks by eliminating many of the flavors and variations that had been added in the prior decade.

The big change, though, was adding new products around its traditional drinks—gels and bars for energy prior to exercise, and protein smoothies and shakes for recovery after exercise. It all added up to an intuitive 1-2-3 G Series of products needed by an athlete before, during, and after exercise for peak performance. Prices rose to reflect this more targeted line of products. Where previously a 32-ounce bottle of Gatorade might have cost $0.99, a 12-ounce bottle of the new before-workout carbohydrate drink sold for $2.99.

Before this huge product transformation, Gatorade sales came almost entirely from mass-market retailers. Now the G Series was also available where serious athletes went—cycling shops, running specialty stores, sporting goods retailers, and even retailers like GNC vitamin shops.

Traditional Gatorade retailers pushed back; this wasn't the business they were used to. Not only were the products and the in-store displays different, but mass-market retailers no longer saw as much Gatorade advertising on television, where previously 90 percent of its advertising budget had gone. No more Super Bowl ads—"Why advertise," Robb O'Hagan asked, "when the players are drinking your product throughout the game?" Now 30 percent of the ad budget went online to social media and niche sites that attracted serious runners and other high-performance athletes.

In light of research that said young athletes start to compete seriously around age eleven, Gatorade returned to working with coaches of adolescent athletes to stress the link between athletic performance and nutrition. The company delivered the same message in lesson plans for its four-thousand-plus sponsored summer camps and sports tournaments. And it began to sponsor training groups that local retailers organized.

Returning to its scientific roots, Gatorade expanded research done at its Gatorade Sports Science Institute in Illinois to understand the physiology of athletic performance and the role of proper nutrition. And it opened a new facility in Florida, where it tested athletes.

In the end, the Gatorade team was able to revive a revered but stumbling brand. By 2015, Gatorade had regained the share lost to Powerade and restarted growth.[3] And it did this in a parent company that, heart and soul, was a beverage company where the highest priority was to restore the Gatorade drink to its glory days.

Some may say that what Robb O'Hagan did was smart but hardly a different way of innovating. Gatorade expanded its product line and thereby grew sales. What's new or different about that?

We understand. What happened at Gatorade can seem obvious at first glance, but look more carefully, and you will see something not readily apparent.

Robb O'Hagan and her team didn't add nutrition products to compensate for declining sales of Gatorade the drink. They expanded the product line for precisely the opposite reason: to grow sales of the sports drink itself, the company's core product. That was their mandate, and that's what she and her people did. When they launched this new approach with all the elements described above, they quickly saw an uptick in sales of the drink (figure 1-1). Sales of the pre- and post-exercise nutrition products followed more slowly as serious athletes began to understand and adopt the Gatorade system.

Common sense might say that to grow a product, you must change or improve the product in ways that make it more appealing. But that's not what happened at Gatorade. Robb O'Hagan and her team updated the product label, cut the number of drink varieties, and focused on a smaller customer segment. But they didn't change the product itself in any meaningful way.

What they did do was innovate around the core product, the drink. They added complementary innovations, so called because the innovations complemented the core product without changing it. Those innovations then worked together and with the core product to make the core more attractive to key customers and thus to increase its sales.

FIGURE 1-1

Gatorade drink sales after innovation around the core product

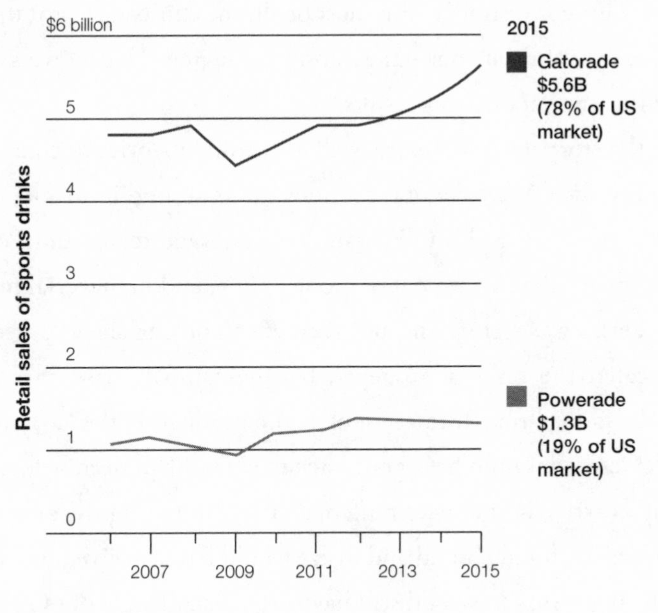

Source: Euromonitor International

This is the first distinctive feature of what we're calling the Third Way. Companies pursuing the Third Way create multiple complementary innovations around a core or key product that make that product more appealing and competitive.

In the Third Way, these complementary innovations possess three characteristics that are exemplified in the Gatorade turnaround. The innovations were diverse, targeted a specific set of customers, and posed little strategic risk.

The steps taken were indeed diverse; they included not just the addition of nutrition products but also a new name and logo, changes in distribution, increased nutrition research, testing of athletes, and programs that allowed serious athletes to better understand their nutritional needs. The target customer, of course, was the serious athlete who had been the Gatorade target customer since the brand was first born.

The feature of "little risk" for Gatorade requires some explanation. As you will see in the company stories to come, complementary innovations around a key product are often "little risks," that is, they require relatively little investment and, most of all, the consequences if they fail would be painful but not dangerously damaging. Hence, we say that little ideas can produce big results.

Yet the story of Gatorade, as well as the other stories to come, seem to test this idea. Were Gatorade's innovations around its drink product really "little"? To Robb O'Hagan, her Gatorade team, and PepsiCo management, the investment of money, time, and managerial focus in the G Series was significant, not little. But nothing she and her team did created the kind of strategic, bet-the-company risk that comes from the radical transformation of a core product. If the G Series had failed, Gatorade could have gone back to what it had been before Robb O'Hagan arrived—not better off, of course, but not significantly worse either. The Gatorade brand and drink would have survived to compete another day. Growth would still have been a challenge, but Gatorade's ability to address it would not have been seriously diminished.

To sum up this important point: most Third Way complementary innovations, when viewed from outside the company, seem relatively little—that is, they entail relatively little strategic risk. But the view from inside the company can be quite different. The investment of money and managerial credibility can be significant, as they were at Gatorade. But, even when this happens, these exceptions never reach the level of strategic risk where failure can be catastrophic for the core product, the brand, or even the company itself. This is a key feature of the Third Way. The core product always remains intact, no matter what happens to the complementary innovations around it.

In our experience, some people, when hearing this story, are likely to say that what Robb O'Hagan and her team did at Gatorade wasn't really innovation, no matter how smart or successful it was, because none of it was truly new. The drink was not new, nor were energy bars or protein shakes.

This reaction is based on a definition of innovation that's fairly common, but that we strongly disagree with: that something is an innovation only if it's disruptive—that is, if it's truly new to the world and changes our behavior in fundamental ways. We believe this definition is too narrow and restrictive to be useful to companies. Instead, we define innovation more broadly, as a new match between a solution and a need that creates value.[4] This broader definition invites a wider range of ideas: an existing technology brought into a new market; an existing need satisfied in a new way; or an ingenious solution redeployed to serve a new group of customers. Innovation, under this definition, includes new ways of combining elements, none of which by itself—need, solution, or value—has to be new. Thus, Gatorade's nutrition products were an innovation in the context of Gatorade's product line, though neither the nutrition products nor an athlete's need for nutrition was new to the world.[5]

We also sometimes hear a second objection. When we talk about disruptive innovation, someone will remind us that we're using *disruptive* in a way that differs from the way it was first used. That's true. We know that Clayton Christensen originally defined the term as referring to new technologies that were less expensive but less capable than existing solutions. These technologies improve rapidly and disrupt the existing solutions, putting the incumbents out of business. Christensen and his colleagues have tried (unsuccessfully) to reclaim their original, more specific meaning of the term.[6] However, it seems obvious to us that *disruptive* and *disruption* have long since passed into broader usage, just as happened with such other terms as *core competence* and *reengineering*. We wrote this book for practicing managers, and among that group, the term *disruptive* has come to mean any innovation that upends existing markets and organizations. As a result, we use the terms *disruptive*, *radical*, and *revolutionary* interchangeably in this book.

In Gatorade, we see the first key characteristic of the Third Way—a set of complementary innovations around a core product that make the product more appealing or valuable. But that's not the only characteristic that differentiates this approach from other types of innovation. To illustrate the other characteristics, we'll now turn to the story of Novo Nordisk and human growth hormone therapy, an example that provides further insight into what sets this approach apart.

Novo Nordisk and the US Market for Human Growth Hormone

In 1997, Novo Nordisk, the multi-billion-dollar global health-care company based in Denmark, introduced Norditropin, its human growth hormone (HGH), to the US market. There the company faced Nutropin, which Genentech had developed and introduced in 1987.[7] Though the two drugs were essentially the same and Genentech had a ten-year first-mover advantage, Novo Nordisk went on to achieve nearly

$1.2 billion in global sales of Norditropin in 2015, owning 32 percent of the market, while Genentech's Nutropin, for years the leader, barely reached $200 million, only 5.7 percent of the market.[8]

Users of HGH are typically adolescents or teenagers of short stature who must inject themselves every day—at best an unpleasant and inconvenient procedure. It's no surprise then that missed doses are a problem. Some 23 percent of teenage patients miss two or more doses a week, which blunts the drug's effectiveness.

The second problem with HGH therapy is the time-consuming, complicated, and often frustrating process every new HGH patient must follow to obtain insurance approval. All involved—patients, parents, pharmacies, doctors, and other health-care professionals—find running this obstacle course exasperating but unavoidable because the drug costs several thousand dollars per month per patient, an amount few families can afford without help.

Perhaps because of its long experience selling insulin to diabetes patients, Novo Nordisk understood from the start the compliance problems raised by a drug that required daily injections. It offered convenient prefilled HGH injection pens that contained multiple doses, made setting the proper dose easy, and required no preparation before using. Competitors offered similar devices, but Novo Nordisk's went further. Its pens were super sharp, which reduced pain and discomfort; they seemed to fit the patient's typically small hands especially well; they wouldn't allow an insufficient dose to be taken; and they were easier to use—the patient pushed a button, and the pen did the rest. Other pens required the user to operate a slider, a more awkward arrangement. Last but not least for a young patient, the Novo Nordisk pen did not require refrigeration after first use. It could be left out with no harm to the HGH.[9]

To address the problem of insurance approval, Novo Nordisk introduced Nordicare, a support program that went beyond anything offered by Genentech or other HGH providers. As soon as a doctor recommended HGH therapy for a patient, Nordicare assigned a case manager who assessed the patient's eligibility and then guided all involved through the approval process. Assistance included advice, tools that helped physicians adhere to prescribing guidelines, reminders of steps to be taken, assistance with appeals, and other kinds of support and encouragement. In addition, Nordicare supplied a starter package for new patients in the approval stage that consisted of a free starter supply of HGH pens (up to three months of pens, if needed), as well as useful paraphernalia: a backpack, a disposal box for used needles, and a pen carrying case for use when the patient was away from home.

Novo Nordisk couldn't improve the product itself. But it could recognize the many problems associated with HGH therapy for all parties and create complementary innovations that solved or relieved those problems. Solving those related problems influenced which HGH therapy doctors chose to prescribe and patients favored. As simple as these steps appear to be, Genentech and other competitors were slow to recognize their importance and respond. By the time they did, Novo Nordisk's Norditropin had become the preferred choice.

Though a drug company and a drink company operate in different worlds, it's not hard to find parallels between the turnaround at Gatorade and the successful introduction of Norditropin. For both, the goal was to extract maximum value from a key product, which is the whole purpose of the Third Way. In both, success was driven by innovation that was neither sustaining nor radical. In both, the companies achieved spectacular product success without much internal disruption and without major changes or improvements in the product itself. Both defied the common assumption that making a product more appealing requires improving the product itself with more features, better performance, new models, and so on. For both Novo Nordisk and Gatorade, the complementary innovations were "little ideas" in the sense, as we explained earlier, that they posed little if any strategic risk to the core product or company. If they hadn't worked, both companies could have returned to where they started.

Both succeeded in the face of serious competitive challenges. Gatorade enjoyed no patent protection or technical advantage. Competitors like Powerade, owned by the Coca-Cola company, constantly undercut it on price, and anyone could enter the market with a new sports drink. Novo Nordisk was ten years late to the US market for HGH therapy, which it entered with a me-too product, and faced an entrenched leader.

And finally, both Gatorade and Norditropin succeeded dramatically and both did it using the same approach. In the Gatorade story, we highlighted the first key feature of the Third Way—a set of complementary innovations around a key product that makes the product more attractive without changing it in any significant way. That's precisely what Novo Nordisk did. It surrounded Norditropin with a multitude of complements that solved or relieved problems related to using HGH therapy.

In Norditropin we can also see a second distinctive characteristic of the Third Way: the complementary innovations operate together and with the key product as a system to carry out a single strategy or purpose—what we call the *promise* to the user. This means the complementary innovations are far more than a random or opportunistic collection of what's convenient or merely possible. In the Third Way, they are aimed instead at collectively satisfying a compelling user need; the promise is a pledge to satisfy that need.

Norditropin and the innovations around it promised to make the whole process of starting and using HGH therapy as trouble-free, foolproof, and pain-free as possible for all involved. Gatorade promised to provide to the serious athlete all the fuel—hydration and nutrition—needed for peak performance.

What can make this second feature less than obvious is that complementary innovations used in the Third Way can look, to a casual observer, like a collection of mere add-ons or obvious choices—low-hanging fruit, as we heard one manager call them. But this feature—focusing the complementary innovations on fulfilling a single compelling promise—is exactly what makes the Third Way such a powerful tool.

CarMax and a Better Way to Sell Used Cars

To define the Third Way further, we turn from corporate turnarounds and successful product launches to a startup—CarMax. This superstore seller of used cars achieved sales of $15.1 billion in 2015, making it by far the biggest and most successful retailer of used cars in the United States.[10] It succeeded in this $200 billion market by using the Third Way to overcome two major obstacles: it could not improve the product it sold, even if it wanted to, and it did not always offer the best or cheapest product available.

Created by Circuit City when the electronics retailer found itself struggling for growth in a saturated domestic market, CarMax opened its first superstore in 1993 and met with great success. It expanded rapidly as it applied its retailing prowess to this new venture in a market that was then dominated by small, local operators—new-car dealers selling used cars, independent used-car lots, and individuals selling a car or two at a time.

A key part of that prowess was the collection and use of data. CarMax put radio frequency identification (RFID) tags on cars and salespeople, registered customers as they came in the door, and tracked what customers looked at, in the computer and on the lot, before they made a purchase (or not). Bar codes added to every car and display space on its lot also let CarMax track sales by lot location and the effects of displaying different kinds of cars next to each other.

Data provided important insights. It helped CarMax determine what to select and pay when buying used cars in the wholesale market. It helped display cars to their greatest advantage on the lot—for example, by showing which display positions sold most quickly or revealing that compact cars displayed next to SUVs looked puny and vulnerable. By providing guidance for repricing and adjusting inventory levels, data also helped the company weather cycles of economic stress or spikes in fuel costs that stifled other retailers' car sales.

Sophisticated skills like these, along with a huge selection of cars to choose from, let CarMax succeed over the next two decades in spite of major hurdles: two painful economic downturns when overall car sales tanked and multiple startup competitors attracted by its early success. Among the competitors was AutoNation, a formidable and better-funded opponent expressly aimed at CarMax's destruction.

What made CarMax's success most impressive, however, was its ability to surmount a life-threatening structural disadvantage. The product it sold—late-model, quality used cars—was a commodity. There was no real difference between a car on its lot and the same make, year, and model on a competitor's lot—except the competitor was often able to acquire and then resell its car at a lower price. The best used cars went at the lowest prices to new-car dealers as trade-ins. Because it didn't sell new cars, CarMax didn't get the great trade-in from the little old lady who only drove to church on Sundays and bought a new car every year or two. It had to acquire most of its product from the more expensive wholesale market.

It's easy to underestimate the gravity of this challenge. If buyers could get the same product at a lower price down the street, why would they buy from CarMax? This was a key reason no one had ever been able to build a large retail chain focused exclusively on used cars. The challenge for CarMax was to find a way of selling a product that was essentially a more expensive commodity, a product it could do nothing to improve.

Here's what it did. It maintained a large inventory of quality used cars. It fixed the price of each car it sold and each trade-in car it bought; there was no haggling. All ancillary elements of the overall deal—the trade-in, the warranty, the financing—were handled separately, and each element came with a no-haggle price as well. CarMax paid the same for a trade-in whether the customer bought a car or not. It took pains to sell only cars in good condition and guaranteed the quality of each with liberal return policies. It paid its salespeople the same commission for every car sold, regardless of the car's price, which removed any incentive to push expensive cars the buyers didn't want. A friendly blue and yellow decor—the same colors used by IKEA—welcomed visitors to every CarMax store. The sales spaces in each store were set up so the customer looked at the same computer screen that the salesperson was viewing—no more looking across a desk at a salesperson who was staring at a screen only he or she could see.

The reason for all these features, none big or disruptive and none that changed the product for sale, came from something the founders of CarMax had discovered when they were first considering the used-car business: most people hated the existing process of buying a used car. Sleazy, dishonest, tricky, and high pressure were some of the descriptors linked to the stereotypical used-car salesman. Perhaps exaggerated, the stereotype nonetheless seemed to reflect many people's experience. In one study that asked respondents to list their favorite and least favorite activities, people on average said they would rather visit the dentist than shop for a used car.

With this insight, CarMax set about creating a buying process that minimized buyer discomfort and distrust, which was not easy to do. To fulfill its promise every day with every customer, the company had to create business systems hardwired to make it happen. For example, to prevent its salespeople from acting like typical "used-car salesmen," it hired salespeople who had never sold used cars before; old habits, it found, were too hard to break. Even then the temptation to haggle and play confusing games with prices was often too strong, especially because each location was responsible for profitability. So CarMax created custom IT systems that separated and fixed the prices for all the different components of a transaction—the price of the car being purchased, the value assigned to the customer's trade-in, and the costs of the ancillary products such as insurance, extended warranty, and financing. The prices and sales process couldn't be changed, even if a salesperson or location manager wanted to.

In addition, CarMax laid out its superstores around a "we're in this together" approach to customers. Its hiring and training systems were designed to find and prepare employees who could carry out its unique approach. In short, virtually all functions and systems within CarMax—not just sales but also compensation, hiring, training, IT, buildings and grounds, inventory management, and more—were consciously and proactively dedicated to fulfilling the promise.

CarMax is yet a third example that doesn't fit today's binary thinking about innovation. Yet pursuing the Third Way enabled it to ward off daunting competition without radical change or internal disruption. Indeed, CarMax used mature technologies in a mature business to succeed in selling the same product that was available, often at a lower price, from its competitors. It succeeded because its promise was so unique and compelling: to make the entire process of buying a quality used car stress-free and transparent, rather than the retail equivalent of a root canal. That's what pulled in customers, brought them back, and made them willing to pay slightly more for the same product they could buy for less from a competitor.

In CarMax we can identify a third characteristic that distinguishes the Third Way: the complementary innovations—even those delivered by outside partners—are closely and centrally managed by CarMax. Though this may seem like a small point, it's crucial for Third Way success. If the complementary innovations must all work together to satisfy a customer need, this will only happen if each innovation can be closely managed.

You can see this feature in the Gatorade and Novo Nordisk stories as well. None of the complementary innovations those companies placed around key products was independent and out of their control. The need for retaining control will become even more obvious in future chapters, where we discuss working with and depending on outside partners who provide important complementary innovations difficult to create inside your own organization.

It's hard to imagine three more-diverse companies: Gatorade, a sports drink; Novo Nordisk, a producer of HGH therapy; and CarMax, a big-box retailer of used cars. The products each sold and the markets in which each operated could hardly be more different. And none, to our knowledge, consciously said, "We're going to innovate in a different way." They simply did what seemed appropriate, given their circumstances. Yet each found its way to essentially the same approach, to what we call the Third Way.

That's the most dramatic endorsement of all. Three different companies (and more, as we'll see in future chapters), operating independently in different industries, found success by innovating in the same low-risk, high-reward way. They all discovered the same insight: that great products may not be enough today. The fortunes of a product can depend less on the product itself than on a group of small, supporting innovations around it.

These three stories—a dramatic corporate turnaround, the successful introduction of a me-too product in an established market, and a startup that was the first-ever business of its kind to succeed—all demonstrate that radical innovation is not the only way a company can innovate today. Indeed, the Third Way is more than a mere option. Sometimes it's the smartest option. Could Gatorade, Novo Nordisk, or CarMax have done better with radical, disruptive change? They all did very well with the Third Way, and none had to take big risks or disrupt itself in the process.

The Three Distinguishing Features of the Third Way

The stories of these three companies illustrate the three key features of the Third Way.

First, and most obvious, the Third Way consists of multiple, diverse innovations around a central product or service that make the product more appealing and competitive. We refer to the product at the center of every Third Way project as the *key* or *core* product. It is always a key or important product; making a marginal product the focus of so much effort would make no sense. But the product does not always have to be a company's core product, as its sports drink was for Gatorade and used cars were for CarMax. For Novo Nordisk, its HGH drug was certainly important, but its insulin product was, at least for the period covered in our story, the company's core product. "Always key and often core" is the way to understand any product that is the focus of the Third Way.

By *diverse* complementary innovations, we mean that they should fall into a wide range of business categories, such as pricing, marketing, operations, sourcing, and partnerships. Likewise, the innovations should appear in a host of different forms, such as auxiliary products, support services, and social media activities.

Second, what makes this approach work is that all the complementary innovations operate together as a system or family to satisfy a compelling promise to the customer. Gatorade promised peak performance for serious athletes through a complete nutrition and hydration solution. Norditropin promised to make HGH therapy as trouble- and pain-free as possible for all involved. And CarMax promised buyers a hassle- and worry-free experience when they were locating and buying the car they needed.

Third, and perhaps the least obvious in the stories, the family of complementary innovations must be closely and centrally managed. It's not an ecosystem of interrelated but autonomous companies and products that compete, collaborate, or otherwise coevolve according to their own needs and priorities. Instead, each complementary innovation is created or selected and then closely managed, usually by the owner of the key product. Indeed, the careful selection and proactive management of this system is crucial to the success of the Third Way.

Is the Third Way Truly a New Approach?

When we've discussed this approach to innovation with managers from around the world, someone at this point usually asks, "Is this really a new approach?" or "How is this approach to innovation different from . . . ?"

The first question is easy to answer: this is not a new approach. All the companies we cite have obviously found their own way to it already. In fact, as we'll discuss in chapter 8, Walt Disney used this strategy back in the 1930s to make his animated motion pictures irresistible to kids. However, to our knowledge, no one until now has explicitly defined and described this form of innovation as a replicable process.

As for the second question—"How is this different from . . . ?"—we encourage any fervent advocates of blue ocean strategy, disruptive innovation, design thinking, lean startup, or any of the many other approaches to innovation that are out there, to read the sidebar "The Third Way and Other Approaches to Innovation" toward the end of this chapter. The sidebar compares the Third Way with such other approaches. We have not hesitated to take ideas from these different approaches (with proper attribution, we hope) and integrate them into our process.

What's new in *The Power of Little Ideas* is not just the *what* but also the *how*. We start by identifying an innovation approach that some companies have already used with great success but that has not been studied or well understood. By studying what these companies have done, comparing the differences between winners and losers in a market, and drawing lessons from these observations, we can provide explicit guidelines for pursuing the Third Way in your organization. An important part of that guidance will focus on how to address the organizational problems and management challenges that make this approach difficult to implement. As far as we know, no one until now has provided such insights.

Innovating around an important product is hardly the right response in every setting to every threat. It wouldn't have saved Kodak from the tsunami of digital photography and the loss of its film business. To survive, that venerable company would truly have had to turn itself upside down and become a camera company rather than a film company.

Where conventional wisdom says radical innovation is the only option, leaders and their organizations owe it to themselves and their stakeholders to explore this less risky option before throwing their organizations into the maelstrom of perilous, radical, disruptive change. And sometimes, as we will show in the next chapter, the Third Way can be combined with other approaches to build something revolutionary. Apple Computer in 2001 launched its Third Way strategy, which laid the groundwork for the truly disruptive innovations that followed.

If your core product is no longer new and your company is simply satisfying the same need that it has satisfied for many years, you should consider this approach. If your company innovated many years ago, but other companies have come into the market with similar ways of satisfying the same need, then your core product may have become a commodity. The Third Way may offer a path out of this situation.

The Third Way is an option to be considered in any setting where the goal, for whatever reason, is to derive maximum value and competitive edge from an important product. Done the right way, its great advantage can break the link between risk and reward. Large rewards can flow from little ideas.

In the chapters ahead we will explore the Third Way in more detail. Chapter 2 retells the stories of LEGO and Apple from a fresh perspective to illustrate the three characteristics that set the Third Way apart from other approaches. Chapter 3 describes the four key decisions that you need to make if you want to pursue this approach, and discusses why and how those decisions pose challenges for many organizations. Chapters 4–7 delve more deeply into how to work through each of those crucial decisions. And, finally, we end with the story of The Walt Disney Company, and how Walt and his brother Roy used a Third Way–like approach starting in the 1930s to keep their studio afloat and to build the company we all know today. Our goal is to provide the necessary insights, guidance, and examples any company can use to put this low-risk, high-reward approach to work.

Three Takeaways for Chapter 1

- The binary view of innovation—that the only alternative to incremental improvement is radical disruption—is dangerously simplistic.
- There is another way that is proven and uniquely different. The controlled development of complementary innovations around a central product, an approach we call the Third Way, can deliver explosive growth without the high cost and risk of radical disruption.
- The Third Way is not a replacement for incremental improvements or disruptive innovations; rather, it's another option that every business leader should understand and consider when faced with an innovation challenge.

COMPARING THE THIRD WAY TO OTHER INNOVATION APPROACHES

After initial exposure to the Third Way and stories of companies like CarMax and Gatorade that have used it successfully, managers often ask how it compares with other approaches to innovation. How is the Third Way different from the blue ocean strategy? How is it different from the lean startup approach? Are they mutually exclusive—that is, must companies choose one or the other? Or can they be combined and pursued together?

In general, the different schools of thought about innovation can be classified into two categories: where to innovate and how to innovate (figure 1-2). The first category focuses on specific strategic choices about which types of markets to choose and

FIGURE 1-2

Two schools of thought about innovation

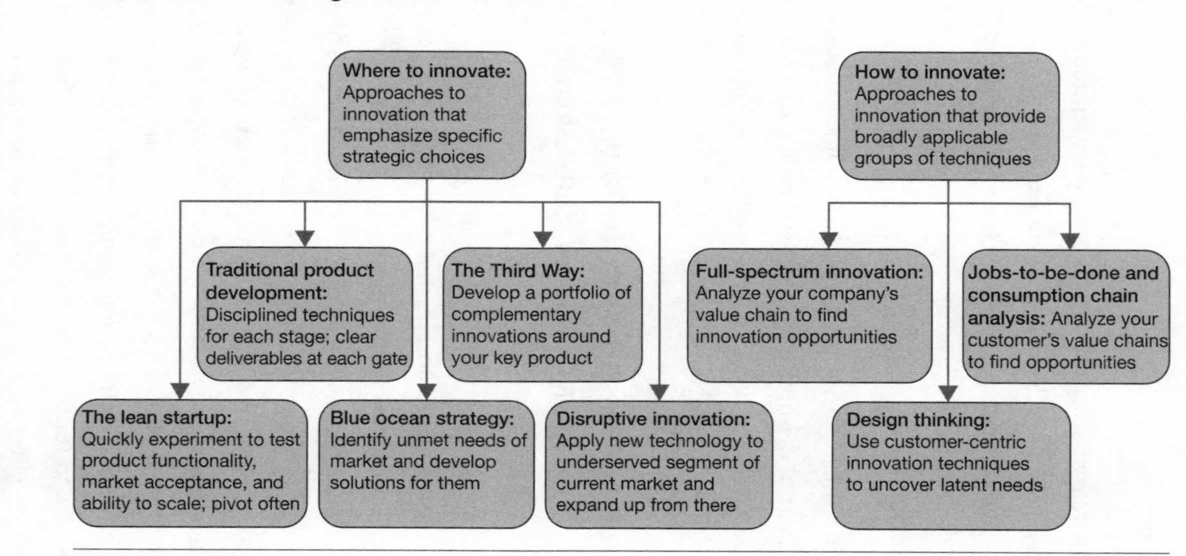

how much technological or business risk to take on. The second category provides bundles of techniques and advice for applying the techniques that can be used for any type of innovation.

In some cases, research groups have created both. The disruptive innovation work by Clayton Christensen and his colleagues (a strategic approach to innovation) spawned some excellent work by Anthony W. Ulwick and others—the jobs-to-be-done framework. These thought leaders provide great techniques for uncovering disruptive opportunities as well as many other types of innovation. The jobs-to-be-done framework is broadly applicable to many types of innovation, and in chapter 5, we show how it can help identify opportunities for the Third Way.

Where to Innovate: Strategic Approaches to Innovation

One of the most frequently used matrices in the innovation literature is the Ansoff matrix (figure 1-3). The two dimensions of the matrix—the degree to which the product or the market represent a familiar or new set of challenges—was first conceived in 1957 by Igor Ansoff in an article titled "Strategies for Diversification."[11] Fifty years later, George Day showed that as a company moves from the lower left to the upper right of the matrix, the probability of failure increases.[12] The following five strategic approaches to innovation can be arrayed on the matrix according to their main area of focus.

The Traditional Product Development Approach

This approach, often called the stage-gate process, is a sequential set of process and review steps that many companies have used for years to develop and improve their products.

FIGURE 1-3

The Ansoff innovation matrix

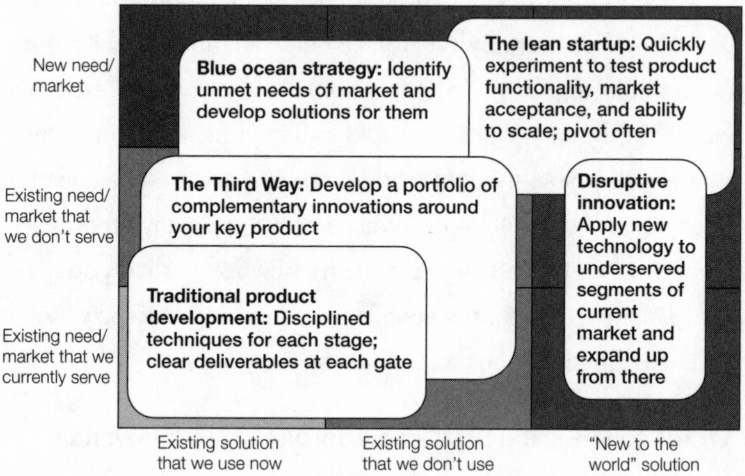

New need/ market

Existing need/ market that we don't serve

Existing need/ market that we currently serve

Blue ocean strategy: Identify unmet needs of market and develop solutions for them

The lean startup: Quickly experiment to test product functionality, market acceptance, and ability to scale; pivot often

The Third Way: Develop a portfolio of complementary innovations around your key product

Disruptive innovation: Apply new technology to underserved segments of current market and expand up from there

Traditional product development: Disciplined techniques for each stage; clear deliverables at each gate

Existing solution that we use now

Existing solution that we don't use

"New to the world" solution

The process starts with customers, suppliers, marketers, engineers, and others contributing ideas for new features or add-ons. Once a company chooses an idea—say, a better alarm clock—it initiates a multistep process of development and review through which the clock is developed and launched into the market. This is sustaining innovation and not the Third Way, which focuses on innovating around the product. There are many excellent texts on this approach to innovation, all of which emphasize disciplined techniques during the stages and the specific deliverables to be reviewed by management at each gate.[13]

Disruptive Innovation

As defined by Christensen and his colleagues, a disruptive innovation is a new product or service that offers both a clear advantage over current products and a distinct disadvantage that prevents acceptance by most current customers. The new product survives as a niche product for a small segment of underserved customers who value its advantages and can accept its disadvantages. Over time, though, the new product's developers find ways to remove its disadvantage. At this point, the new product enjoys rapid acceptance by most current customers, who abandon the old product, and the new product developers displace the old market leaders. As we described in this chapter, disruptive innovation is the form of innovation so much in vogue today and whose proponents are advising current market leaders to "disrupt yourself before someone else does."

We, of course, present the Third Way as an alternative that companies should explore before they make radical, risky changes in their products, processes, and business models. But as we'll discuss in chapter 2, the Third Way is not incompatible with a disruptive approach; Steve Jobs used the Third Way to build a powerful system of innovations around the Mac, iTunes, and iPod, and this system helped Apple's disruptive iPhone succeed.

Blue Ocean Strategy

According to this approach, winning companies are those that find unserved or underserved markets where there's no competition (blue oceans), then innovate to satisfy the needs of those markets. W. Chan Kim and Renée Mauborgne's blue ocean strategy laid out a method to create what they called "value innovation": understanding the dimensions that customers care about, mapping out the existing offerings across those dimensions, then figuring out what could be added, removed, reduced, or increased from the offer.[14] By doing so, a company can differentiate itself and "make competition irrelevant."

Blue ocean strategy can be thought of as the opposite of disruptive innovation. Where disruptive innovation is using a new technology or business model to revolutionize an existing market, blue ocean strategy uses existing technology to create a new market.[15]

For companies seeking to develop a blue ocean strategy, the Third Way can be a useful technique for satisfying the unmet needs of an uncontested market. GoPro's action cameras, a story we'll discuss later in more detail, was a classic blue ocean strategy. GoPro's rugged, waterproof video cameras satisfied a latent need in the market, but the company's rapid growth and ability to fight off competitors occurred because of its ability to execute the Third Way. GoPro not only created the action camera market but also cemented its position in that market with a full portfolio of complementary innovations.

Lean Startup Thinking

The lean startup approach suggests that you innovate by adopting the method used by startups trying to bring a new technology or product to market.[16] The goal is to find both a market that wants the new product and a scalable, profitable business model for delivering it. To find this market and business model, startups conduct multiple experiments. If one market and model doesn't work, they pivot to another and then, if necessary, another and another until they find a group of customers who want what they have.

This approach and the Third Way are quite different. Lean startup thinking applies to a new product or technology in search of a market, while our approach is aimed at making an existing product in an existing market more appealing. However, techniques from each approach can be useful in the other; experimentation can play an important role in the Third Way (a topic we'll discuss in chapter 6), and complementary innovations may make a new technology more useful and appealing. But the two approaches are fundamentally intended for use in different settings.

How to Innovate: Innovation Techniques

Design Thinking

Tim Brown's book *Change by Design* and others that followed laid out the argument and a detailed process for this approach, which is distinguished by an intense focus on the user, on investigation of both typical and extreme users to gain insights about user needs, on disciplined observation, and on constant experimentation.[17] The heart of design thinking is to focus on why and how a product or service is actually used by people—customer-centric design. It calls for not just asking customers what they want but thoroughly understanding their lives and the challenges they face.

Organizations pursuing this approach tend to focus on improving the product itself, but they may find the Third Way an attractive option for satisfying the customer needs uncovered by design thinking. Conversely, some of design thinking's tools and processes can help those pursuing the Third Way understand intimately the human context in which a key product is used. Such information can reveal not only a compelling promise but also complementary innovations that satisfy that promise.

Innovating Your Company's Value Chain: Full-Spectrum Innovation

In 1985, Michael Porter published his masterpiece, *Competitive Strategy*, in which he laid out his view of the organization as a linked chain of activities, what he called a "value chain."[18] Since then, many have used this view of the organization as a way of analyzing what a company does and finding opportunities to improve those activities.

The most direct descendant of Porter is the full-spectrum school of innovation. Whether it's called full-spectrum innovation or the ten types of innovation, this approach emphasizes the importance of innovating in all parts of the company and with external partners.[19] Like the Third Way, these approaches are based on the notion that great products are not enough. To succeed, firms must also innovate in the way they manufacture, distribute, sell, and support those products.

The similarities between full-spectrum innovation and the Third Way are obvious, but what sets the Third Way apart is whose value chain you innovate around. Where the full spectrum focuses on how to innovate along the value chain of the company producing the product, the Third Way focuses on how to innovate around both the producing company's value chain and the customer's value chain. This focus leads to a very different set of techniques and recommendations than does a full-spectrum approach. It also leads us to spend much more time addressing the organizational and managerial challenges raised by this kind of innovation, a topic little covered in the literature if at all.

Innovating the Customer's Value Chain: Jobs-to-Be-Done Analysis and Consumption Chain Analysis

A second stream of work also uses this Porter-like value chain approach, but applies it to the customer's business. There are two schools of thought on how to analyze the customer's value chain: jobs-to-be-done analysis and consumption chain analysis.[20] Jobs-to-be-done analysis is based on the simple but powerful idea that customers don't buy a product because they want the product; they want a job done, and they buy a product or products because they want to complete that job. In other words, no one buys a drill because they want a drill; they buy it because they want a hole. Proponents of this approach claim that most products force the customer to accept a compromise; that is, most products fall short of doing the actual job customers want done. Consequently, the frustrated customers are then forced to assemble the full solution themselves. Advocates have constructed around this notion a variety of processes, tools, and diagnostics to help identify and satisfy the difficult or frustrating steps in the process.

Consumption chain analysis also studies the customer's process, but focuses on the parts of the process where money changes hands. For example, where jobs-to-be-done analysis focuses on the preparation, use, and cleanup after drilling a hole, consumption chain analysis focuses on the process of buying a drill, buying drill bits, paying for the power to operate the drill, servicing the drill as needed, and replacing it if it breaks.

Both techniques are entirely compatible with the Third Way. Both map the customer's process and find gaps that are opportunities for a firm either to improve its current product or to find complementary products or services that will help improve the customer experience. Both can help companies pursuing the Third Way understand what users of a key product really want—the job that the product is supposed to do or the process that they have to execute to purchase the product—and then to find complementary innovations to help fill any gap between product and process.

We'll end this brief review by emphasizing a point we made earlier: the Third Way is not the only way to innovate—it's another tool that every manager needs to have in their innovation toolkit. It's not the best approach when you're trying to apply or defend against a groundbreaking new technology—it wouldn't have saved Kodak from the digital photography revolution. But it's an approach every company should consider as it evaluates its alternatives for investments in innovation.

2

LEGO and Apple Computer

Masters of the Third Way

In chapter 1, we saw how three different companies in three different industries found and pursued an approach to innovation. Using the Third Way, they turned around a faltering brand, displaced an entrenched market leader, and launched a successful startup. In every case, they succeeded despite major competitive disadvantages in their core product.

None of these three firms—Gatorade, Novo Nordisk, and CarMax—was following a playbook. Each was simply doing something it considered smart. Yet they all found their way to the same approach. They succeeded by surrounding a core product with small complementary innovations that made the core product irresistible.

Clearly, the Third Way is a powerful tool for extracting maximum value from an important product, without changing the product itself significantly and without taking on major risk. But those three stories don't completely communicate the power of this approach.

In this chapter, we will look at two iconic companies that used the Third Way not to rescue a product but to rescue the company itself when its continued existence was in doubt. We believe that for Apple and LEGO in the early 2000s, the Third Way produced two of the greatest turnarounds in modern business history. We look at these stories because the paths taken by the two companies were remarkably similar and will help illustrate the path any company can take to execute the Third Way.

The Turnaround at LEGO

We begin with the story of LEGO. You may be familiar with what LEGO did to recover from its brush with bankruptcy in 2003 and become the world's largest and most profitable toy company.[1] We return to the story to explain how LEGO ultimately succeeded.

We pick up the narrative in 1993 when the Danish company suddenly stopped growing. It had just finished a fifteen-year period (1978–1993) during which sales grew at 14 percent per year, driven by wave after wave of new construction toys based on LEGO's signature plastic brick. But a flattening of sales starting in 1993 marked the end of an era for LEGO and the beginning of an existential crisis that would last well into the twenty-first century (figure 2-1).

A changing world in the 1990s presented LEGO with a perfect storm of competitive threats and market shifts. Its patents had expired and other toymakers were selling cheaper versions of its signature snap-together bricks. Digital games such as PlayStation and Nintendo were competing for kids' playtime. And while other toymakers were shifting production to China, LEGO molded its plastic bricks in high-cost Denmark.

FIGURE 2-1

LEGO sales, 1978–1998 (DKK billions)

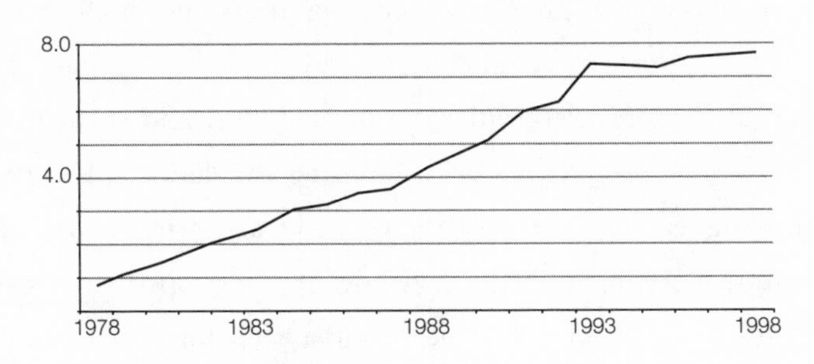

Source: The LEGO Group Financial Statements

A Classic Case of Binary Thinking

LEGO's response was a classic example of the binary thinking about innovation that we have described. First, it reacted the way most companies react to threats and changes—it did more of what had worked in the past. It tripled the number of new toys it brought to market each year, from 109 in 1994 to 347 in 1998. As a result, complexity in the factories exploded and costs skyrocketed. But sales stayed flat and profits dropped, resulting in the first loss in company history in 1998.

After a year of soul-searching and a change in top management, LEGO decided it had to try something else. It was clear to management that incremental innovations weren't enough. Kids were moving to new kinds of play. The era of the brick was over. The company believed it had to put itself through wrenching, uncertain, and risky change if it hoped to survive. This was the period when the idea of disruption and disruptive innovation was beginning to grip the business world, and LEGO was an early convert to the gospel of disruption.[2]

So the second way LEGO responded to a changing world was by innovating away from the brick. Beginning in 1999, it produced a stream of diverse new products, some big and some small, many of which weren't brick-based and didn't require construction. It created LEGO Explore (electronic toys for toddlers) and spent massive amounts developing a virtual brick-building simulation called LEGO Digital Designer. It invested in theme parks and after-school education centers. It commissioned a TV show and developed video games. And taking a direction it had never pursued before and that many in the company resisted, it produced lines of toys around two blockbuster movie franchises, Star Wars and Harry Potter.

Sales picked up for a few years until 2003, when they plummeted. Four Star Wars and Harry Potter movies released between 1999 and 2002 had boosted sales of LEGO toys tied to those films. But no new movies from either franchise appeared in 2003 or the first half of 2004, and sales of those toy lines nosedived (figure 2-2).

It was as if the tide had gone out and exposed a broad expanse of empty, rocky beach. The unhappy truth, now revealed and undeniable, was that the 1999–2002 surge in sales from movie-themed toys had

FIGURE 2-2

LEGO sales, 1978–2004 (DKK billions)

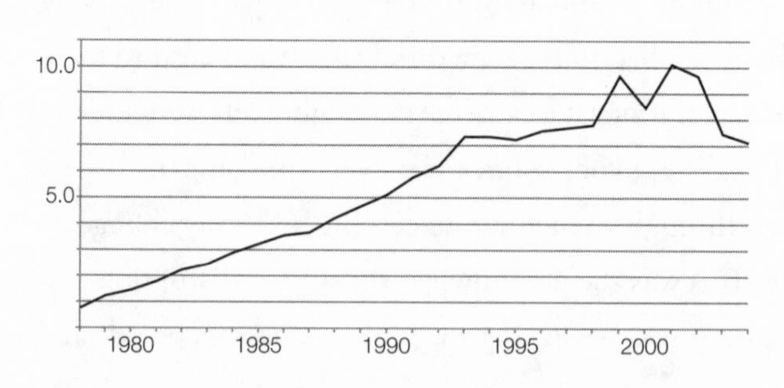

Source: The LEGO Group financial statements

hidden a harsh reality: LEGO had virtually nothing else to drive sales and profits, despite all the radical innovation it had tried. Most of its other toy lines were generating massive losses. The company was running out of cash and had no committed lines of credit. So dire was the situation that a LEGO executive, Jørgen Vig Knudstorp, told his colleagues in 2003, "We're on a burning platform." After hearing Knudstorp's analysis, LEGO hired a new CFO, Jesper Oveson, who reviewed the data and promptly declared the situation "a financial disaster."[3]

LEGO's declining sales exposed the dark side of disruption, the side its fervent advocates didn't discuss much. No one doubted that radical change had become more common and that every company constantly had to search for opportunities to change its world. Yet the downside of disruption was that it failed more often than it succeeded. Incrementally improving current products didn't fail as often or cost as much. But for LEGO, expensive failures were the norm for its disruptive innovation efforts. Its attempts to create revolutionary change almost pushed it into bankruptcy.

LEGO's experience also illustrates a key reason that radical innovation can fail: there are limits on what customers will accept from a company they know. When LEGO created new play experiences without bricks and construction, the change confused the typical toy buyer and infuriated longtime LEGO fans everywhere. The CEO of Toys "R" Us told the company "We love and understand the LEGO brand better than you do." Howard Roffman, the head of licensing for Lucasfilm, offered a similar appraisal: "You've lost your grip on the business. You're not on top of your game."[4] Fans echoed those sentiments by buying other toys. Like many companies, LEGO was associated with a few central products, and when it moved away from them, customers had no reason to do business with the company.

LEGO's experience again raises the question: if incremental improvements aren't enough, is the only alternative radical innovation? Is big, fundamental, risky change the only effective response to a competitive threat? The experience of LEGO—like Gatorade—suggests otherwise. As it scrambled back from the brink of bankruptcy in 2003, LEGO stumbled onto the Third Way.

Saved by Bionicle

In 2003, when the company examined the debris of its failed experiments, it discovered one consistent success—a quirky construction toy called Bionicle. Like other LEGO toys, Bionicle was a set of plastic pieces with assembly instructions. But Bionicle differed in three key ways from what the company had done before. The plastic pieces were used to construct action figures, a new type of construction toy for LEGO. Second, the toy came with a LEGO-created story of heroes battling villains to save the world, again a first for LEGO. And finally, its boxes of plastic pieces were surrounded by a swarm of complementary innovations—new packaging, comics, books, a video game, direct-to-video movies, and a wide array of licensed merchandise, such as McDonald's Happy Meals and Nike shoes.

None of these related innovations was risky or expensive, or even noteworthy by itself, and none could have succeeded alone. But each made Bionicle even more appealing to the target audience: boys between six and nine years old. Even better, LEGO owned, controlled, and profited from all the pieces, unlike what it got from its tie-ins with movies it didn't own. Bionicle marked the first time that LEGO had created a toy around its own intellectual property—a story filled with compelling characters that LEGO had developed itself and owned entirely.

Bionicle was literally the toy that saved LEGO. The company sold over 190 million of the Bionicle figures over the toy's nine-year life. The profits from those sales were the only bright spot during the crisis years of 2003 and 2004. In fact, Bionicle was so successful that over the next four years, LEGO reorganized its development organization to repeat the innovation approach that had made this toy line so successful—compelling story-driven construction play sets, surrounded by low-risk complementary innovations, brought to market by both LEGO and its external partners.

It's easy to see the three characteristics of the Third Way at work in what LEGO did. First, the company surrounded each toy with the kind of complementary innovations we noted for Bionicle, none of which changed the toy itself but all of which made the toy irresistible to kids. Second, all those complementary innovations worked together and with the toy to fulfill the promise of a rich and compelling story-based play experience. And third, LEGO made sure to own or retain creative control over all of the complementary innovations.

Results have been dramatic. In a global market with intense competition, no barriers to entry, fickle customers (whose tastes change faster than a seven-year-old's?), and no patent protection, the approach LEGO discovered has produced an average annual sales growth of 21 percent and average annual profit growth of 36 percent from 2007 to 2015 (figure 2-3). LEGO is yet another company—like Gatorade, Novo Nordisk, and CarMax—where the Third Way has produced great success.

FIGURE 2-3

LEGO sales, 1980–2015 (DKK billions)

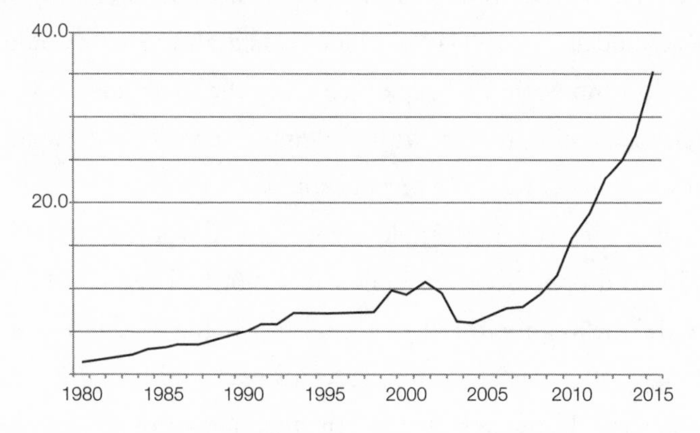

Source: The LEGO Group financial statements

It's instructive to compare the story of LEGO with that of an even better-known but broadly misunderstood turnaround—the story of Apple after Steve Jobs returned in 1997. This is not the Apple story you typically hear—the one that portrays Jobs and Apple as an example of success through radical innovation. In fact, Jobs and Apple initially prospered not by disruption but by the controlled development of complementary innovations around a central product.

The Rise of Apple Computer After 1997

When Jobs returned to Apple Computer in the middle of 1997, he took on a huge challenge. Apple's share of the PC market, 12 percent in the mid-1990s, had just dropped to 4.6 percent.[5] With the release of Windows 95, Microsoft had finally removed much of Apple's better-interface advantage. And Apple's decision to license the Mac operating system to makers of Mac clones was cutting into its own sales of machines without increasing the Mac operating system market share. The company had gone through three CEOs after Jobs was ousted in 1985 and had responded to its competitive woes in classic fashion—with a proliferation of new products and models, including over one hundred new Mac models in 1995–1996 alone. It had also experimented with more radical innovation strategies, entering the PDA market with the Newton, the gaming market with the Pippin console, and digital photography with its QuickTake line of cameras.

Neither strategy worked. Apple Computer's sales peaked at $11 billion in 1995 and dropped to $9.8 billion in 1996. In January 1997, then-CEO Gil Amelio announced that sales in the fourth quarter of 1996 had plunged 30 percent from the previous year, with a market share of just 7.4 percent. In the next quarter, the first quarter of 1997, Apple lost $700 million.

Finally, in July 1997, the board asked Jobs to return, and he agreed. Around this time, someone asked Michael Dell what he would do if he were CEO of Apple. His answer: liquidate the company and return the money to shareholders.

Jobs's Early Moves: 1997–2000

Jobs spent his first few years back at Apple making basic moves to stop the bleeding and get the company back on track. After a quick but intensive review, he dropped 70 percent of the company's hardware and software products.[6] He reduced the confusing multitude of Apple computers to just four: two low-end consumer models (the iMac desktop and the iBook laptop) and two high-end professional models (the PowerMac G3 desktop and the PowerBook G3 laptop).[7] He killed the Newton PDA, the Pippin gaming platform, and the QuickTake line of cameras, and he ended all licenses to produce Mac clones.[8]

He lowered costs by outsourcing manufacturing. He reformed the Apple board by replacing some members with friendlier faces. He brought in some senior people from NeXT, the innovative but unsuccessful computer maker he'd founded after leaving Apple in 1985.[9] He reorganized Apple to do away with competing product domains. And he ended Apple's years-long dispute with Microsoft by striking a deal with Bill Gates that terminated all patent litigation between the two companies.[10] This agreement was widely interpreted to mean that Apple did not plan to compete in the business market.

It wasn't all a matter of clearing the decks. Jobs also initiated work to improve Apple's few surviving products. In 1998, the company introduced the curvy and colorful iMac, which, by its sleek design, announced, "Apple's back!" Perhaps even more important, Jobs initiated a complete redesign of Apple's operating system, based on porting the NeXT operating system to the Mac platform. The change first appeared as the Mac OS X in 2000 but would take until the mid-2000s to unfold completely. It would ultimately serve as a far more powerful and flexible operating system for all Apple products.

The immediate results of these initial moves were only partly encouraging. Apple lost over $1 billion in 1997 but did manage a small profit in the fourth quarter of that year and was profitable in both 1999 and 2000 (figure 2-4).[11] However, sales continued to drop: from $9.8 billion in 1996 to $7.1 billion in 1997 and then to $5.9 billion and $6.1 billion in 1998 and 1999 (figure 2-5). During Jobs' early years as CEO, Apple's PC market share continued to drop—to a low of 2 percent in 2004 (figure 2-6).[12]

FIGURE 2-4

Apple earnings before taxes, 1995–2004

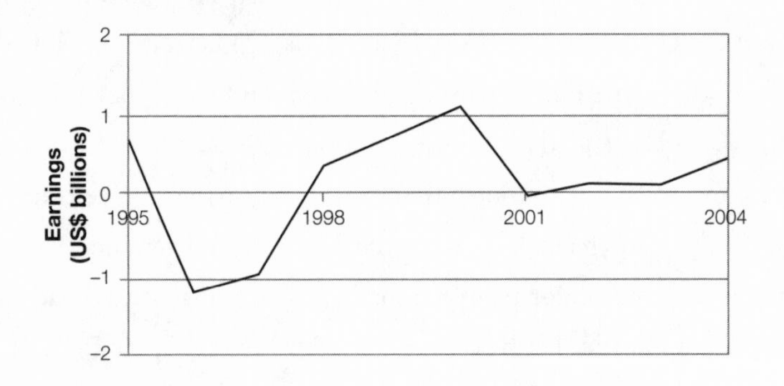

Source: Apple Computer SEC filing

FIGURE 2-5

Apple revenue, 1995–2004

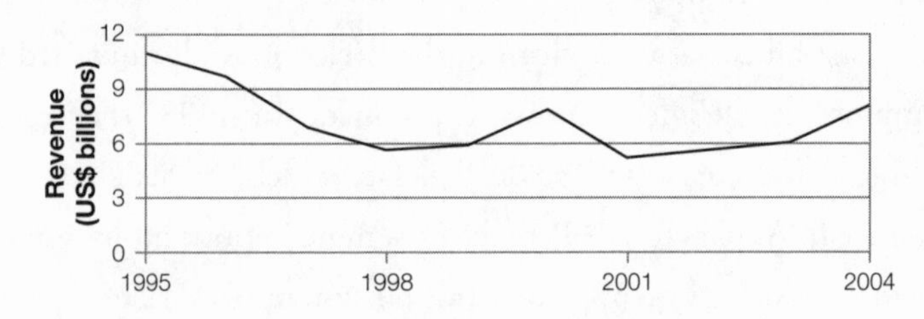

Source: Apple Computer SEC filing

FIGURE 2-6

Apple market share in the United States, 1995–2004

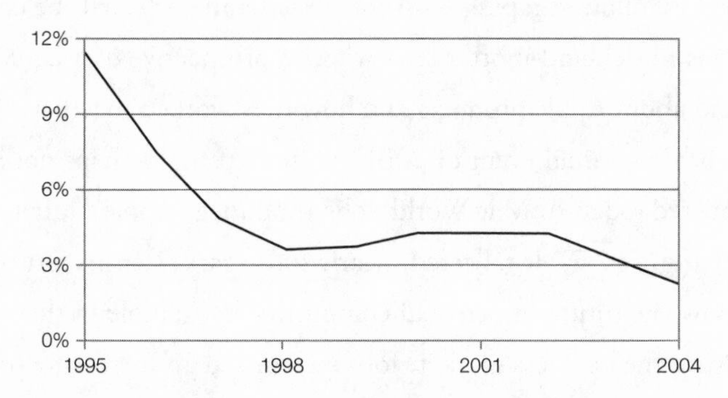

Source: Jeremy Reimer. "Total share: 30 years of personal computer market share figures," *Ars Technica*, Dec 15, 2005, http://arstechnica.com/features/2005/12/total-share/10/.

A Seminal Year: 2001

The year 2001 was the first time that Jobs announced his complementary innovations around the Mac: the iPod, iTunes, and Apple Stores. The iPod was a nicely designed MP3 player with a sleek case, elegant click wheel for selecting songs, and tiny but capacious hard disk that let users store and play not some but all of their songs. It synced effortlessly with iTunes on the Mac and let users "rip, mix, and burn" their music—that is, pull songs from CDs (or piracy sites like Napster), organize them in playlists, and download them to a new CD or the iPod for playing anytime, anywhere.[13]

Neither the iPod nor iTunes was the first of its kind. There were many competing MP3 music players and MP3 management software products. In fact, Apple didn't develop iTunes from scratch. In 2000, it purchased SoundJam, a market-leading application for storing and organizing digital music, brought its developers (two former Apple software engineers) into the company, and relaunched the software as iTunes the following year.

It was also in 2001 that Apple launched its Apple Stores. After months of preparation, the first opened in Tysons Corner, outside Washington, DC; others followed rapidly as Apple tested and perfected the concept. These were brick-and-mortar sites where a prospective user could learn firsthand about Apple products and how they worked together.

The other seminal event of 2001 was something perhaps not as well remembered today. At Mac World, Jobs announced Apple's "digital hub" strategy (figure 2-7). It reflected, nearly four years after his return, how Apple saw the future of personal computing and its role in that future. Eschewing the business market, Jobs recognized an increasing problem for individual users. As more and more parts of their lives were being

FIGURE 2-7

Apple's digital hub, circa 2001

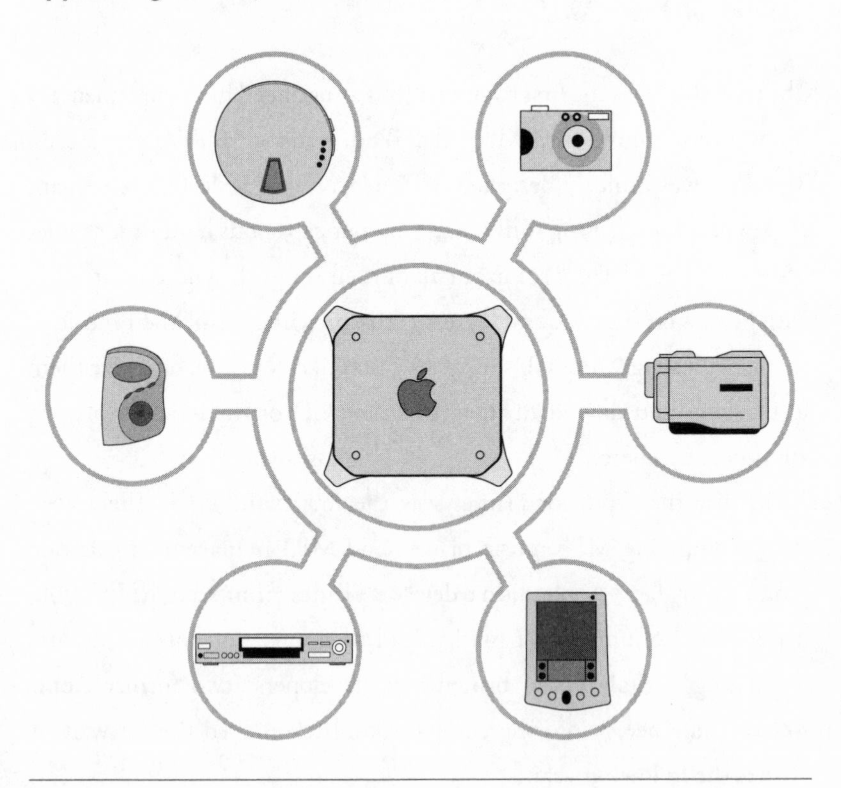

digitized—address books, calendars, photos, to-do lists, feature movies, home movies, music, and more—users needed some way to manage all of it. His basic concept was that Apple Computer would provide the means to do that. In his Mac World presentation, he represented the digital hub with a diagram that showed an iMac surrounded by a VCR, a CD player, a camera, and a PDA. He literally showed the iMac as a central product surrounded by a range of complementary innovations. His message to the crowd? That the iMac could be the control center of the user's digital life.

It was a strategy Apple had already begun to follow. In 1999, iMovie let users shoot home movies on their digital camera and edit them on a Mac. Other realms of digital life followed: iDVD and iTunes in 2001, iPhoto in 2002, GarageBand (for recording and editing music) in 2004, and iWeb for creating websites in 2006. Apple labeled these and other applications iApps (later, iLife) and promoted them as "Microsoft Office for the Rest of Your Life."

All these apps were meant to draw users to the Mac. In fact, Jobs apparently felt iMovie would be the first killer app for the digital hub strategy, but he overestimated people's desire to shoot and edit their own movies. iTunes, of course, was that killer app. Until iTunes and the iPod, digital music was difficult to manage—to get it, download it, mix it, load it on a device to carry around, and play it. The whole experience reflected the disparate needs of the many vendors involved, not the needs of the users who somehow had to make everything work together. The iMac, iTunes, and iPod made it easy. It was the perfect example of what the hub strategy was meant to do, and the Apple Stores were a perfect way to show users how easy these products were to use when attached to the Mac. And around this time Apple ran a television campaign called "Switchers," in which consumers explained how they had switched to the Mac and how it was helping them manage their digital lives.

It's easy to forget, since we know what happened subsequently, that initial reactions to all this were mixed. Apple's digital hub strategy cut against the grain of thinking at that time. With the rise of the internet, conventional wisdom said the PC would soon become little more than a terminal for accessing the online world. Yet here were Jobs and Apple saying the PC, or at least the Mac, would continue to play a key role by itself as a tool for pulling together all the digital pieces of a user's life.

While the iPod was recognized for its sleek design, elegant interface, and high capacity, at $399, it also cost far more than competitive players.[14] And it initially only worked with the Mac, which was the point, of course, but the Mac market share was only some 5 percent of PC users. Some of the less-than-enthusiastic online comments included these: "[Is Apple] really aiming to become a glorified consumer gimmicks firm?" "Great just what the world needs, another freaking MP3 player. Go Steve!" "I want something new!" lamented another, "I want them to think differently!"[15]

Nor were the Apple Stores considered an obvious move, especially at a time when brick-and-mortar was considered the old, dying approach to retailing. Dell, with its direct-to-the-user model, was the new paradigm. Other computer companies, such as Gateway, were leaving the retail arena by 2004, and IBM had left retail in 1986.[16]

While Apple's moves between 1997 and 2003 were very similar to LEGO's, the results weren't as positive.[17] It didn't help that the tech bubble burst in 2000, and so just surviving in the difficult years between 2000 and 2003 was an impressive feat.

The iTunes Music Store: 2003

The year 2003 brought another seminal event, the launch of the iTunes Store, where iPod, iTunes, and iMac users could purchase and download music. In fact, the store comprised a number of innovations that gave Apple users legal access to a vast library of music available not only by album but also by individual song for $0.99 per song. Many would argue that the store was Jobs's first genuinely revolutionary innovation since his return in 1997; we believe otherwise.* But what's indisputable is that Apple Computer would never have been able to negotiate its new business model without the groundwork that was laid in the years before. Music publishers came to the table to negotiate because they were terrified of losing their businesses to illegal file-sharing sites and because their own attempts to sell music online had failed. In that context, Apple's offer was compelling. "You can sign up with us for ninety-nine cents per song," it told the music companies, in effect, "or you can let us continue to encourage people to use unprotected MP3 formats, in which case you'll have to sue your own customers to stop them from copying their own music." What also made this deal attractive was the FairPlay copy protection software Apple had developed. Unlike most other such software, FairPlay was both secure enough to satisfy music companies and unobtrusive enough to satisfy users.

Jobs launched the iTunes Music Store in April 2003, and in the following eight weeks, users bought five million songs.[18] After nearly six months, the number had risen to thirteen million songs, giving Apple 70 percent of all legal music downloads.[19] In October 2003, Jobs launched iTunes for Windows, which allowed Windows users to sync their iPods and download music from the iTunes Music Store.[20] Apple touted it as "The best Windows app ever written."[21] Now the iPod and its many complements were available to virtually anyone with a computer.

*Was it really revolutionary to give people the ability to buy music online in 2003? We don't think so. The first online music service, ritmoteca.com, was launched in 1998. Sony, EMI, and an independent company called eMusic sold music online starting in 2000 and let customers purchase music either by the song or by the album. For a review of these services and their problems, see Richard Menta, "Priced to Lose? Labels Sell Music Online," *MP3 Newswire*, August 3, 2000, www.mp3newswire.net/stories/2000/lose.html.

Also in 2003, people began to notice that Apple was doing something truly different. With the iTunes Music Store an immediate success, one commentator said the iPod was "becoming a cult object, a music player so successful that it embodied the digital music era all by itself." The Apple retail stores were popular, and the company was growing their number rapidly. And while the Mac's market share was still around 2 percent in January 2004, the iPod already accounted for 31 percent of all MP3 players sold.[22]

Still, Apple Computer's sales in 2003 only rose to $6.2 billion (versus $5.7 billion in 2002), and its profitability was only slightly better than 1 percent of sales. The capital markets weren't impressed, and the company's stock price remained low well into 2005.*

Little Ideas Become Big Ideas: 2004–2007

The iTunes Store in 2005 expanded beyond music to eventually offer TV shows, movies, and audiobooks; in short, the store became the source of not just music but many other forms of digitized entertainment as well. That same year, Apple completed its transition to OS X with the release of OS X 10.4.[23] Among other features, the new operating system allowed the move Jobs had announced that summer: Apple was switch-

*Todd Zenger, in his book *Beyond Competitive Advantage: How to Solve the Puzzle of Sustaining Growth While Creating Value* (Boston: Harvard Business Review Press, 2016), argues that strategies, like Apple Computer's, that require analysis by many different financial analysts are undervalued by capital markets. This happens because the analysts for different industries are separated in their firms and unable to correctly value the overall benefit of a full portfolio of innovations. In other words, the capital markets treated Apple as four separate uncoordinated companies—a PC manufacturer, a music seller (iTunes), a consumer electronics company (the iPod), and a retailer (the Apple Stores). Sure enough, even after Apple put all the pieces in place in 2003 and 2004, the capital markets remained unimpressed. This leads to an interesting question: are companies that use the Third Way undervalued by the capital markets?

ing from PowerPC chips to Intel chips, which meant Windows users could switch to the Mac and still run their Windows software.[24]

With 2007 came two more huge developments: a new iPod and the iPhone. The new iPod was completely redesigned. Gone was the click wheel, replaced by the touchscreen with icons we all know today. The iPhone, which was the redesigned iPod with a phone, was now a powerful and flexible mobile device that could satisfy most of a user's digital needs: telephony, online access, calendar, email, contacts, music, movies, TV shows, and a multitude more, depending on what capabilities the user chose to add. In addition to covering all those user needs, it also provided a highly intuitive graphic interface that offered probably the best user experience available at the time. The iPod and iPhone demonstrated Apple's unique skills in both hardware and software and in melding the two into a seamless experience. Tellingly, Apple dropped the word "Computer" from the company's name in 2007.

Apple's sales finally began to take off. The year 2005 was the first year—eight years after Jobs's return—that sales surpassed their previous high of $11.1 billion in 1995 (figures 2-8 and 2-9). Of course, after 2005, fueled by the launch of the iPhone in 2007, the App Store in 2008, and the iPad in 2010, Apple's sales grew dramatically.

FIGURE 2-8

Apple revenue, 1995–2015

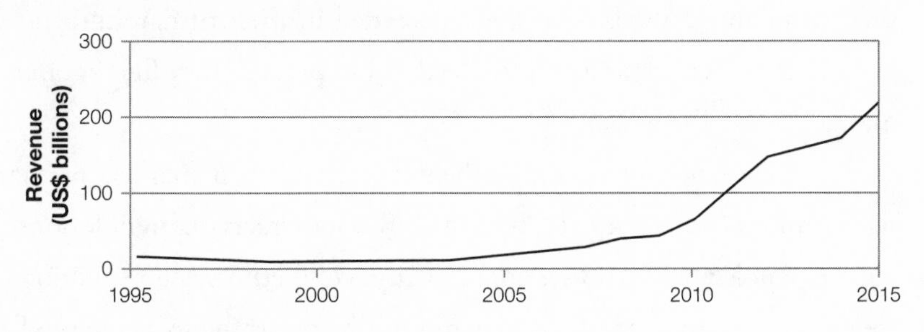

Source: Apple Computer and Apple SEC filings

FIGURE 2-9

Apple earnings before taxes, 1995–2015

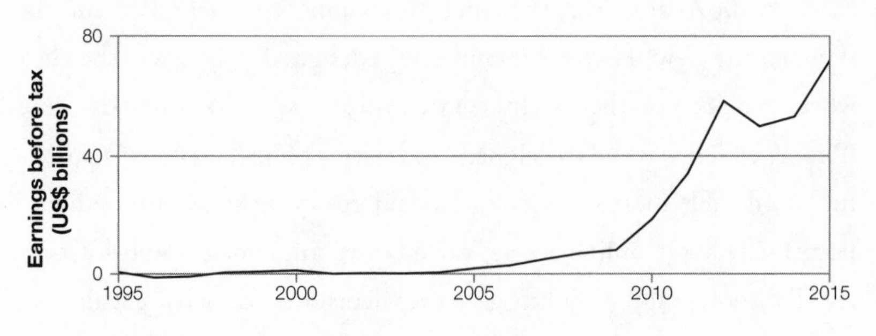

Source: Apple Computer and Apple SEC filings

The Lessons of Apple and LEGO

It's easy to see what Steve Jobs did at Apple as a succession of brilliant game-changing, disruptive innovations, with the iPod, iTunes, the iPhone, and the iPad being the milestones that demonstrated his exceptional leadership. With LEGO, many have pointed to the company's mastery of digital play experiences and computer-generated animation as evidence of its return to innovation leadership in the toy industry.

But focusing on the markets these companies have changed and the ways they've transformed our lives misses an important lesson. The key question is this: How did Jobs and Knudstorp (the LEGO CEO) lead their companies? We believe they succeeded in their turnarounds not because they were great disruptive leaders but because they first became masters of the Third Way.

This view does not diminish their impact on their markets or the significance of what they accomplished. But it extracts different lessons. Neither Apple nor LEGO set out to disrupt competitors and revolutionize markets. Both began their turnarounds by returning to a core product, surrounding it with a portfolio of complementary innovations, and designing those complements to make their core products more desirable and valuable. When—and only when—that system was in place both companies used the financial strength and the credibility they had established to expand into very different markets (feature-length movies for LEGO and phones for Apple) that, at least for Apple, were truly disruptive.

We feature the Apple and LEGO turnaround stories for three reasons: First, they reinforce the three characteristics of the Third Way that we discussed in chapter 1. Second, the stories provide a roadmap for companies looking to apply the Third Way to their own products and services. The paths Apple and LEGO took in their turnarounds were remarkably similar, and these similarities suggest some paths that other companies can follow. Finally, each story shows where the Third Way can lead—that investing in a family of little innovations around a central product can sometimes lead to much larger opportunities in the future.

Both Used the Third Way to Turn Around Their Companies

Let's review the three characteristics of the Third Way approach to innovation.

MANY DIVERSE INNOVATIONS AROUND A KEY PRODUCT. This feature describes precisely what Jobs did once he finished rationalizing the Apple product line and improving the Mac. His goal then was to sell more Macs, and from that point, in 2001, his moves—the iPod, iTunes, the iTunes Music Store with the $0.99-per-song business model, the FairPlay copy protection system, and even the Apple Stores—all served to make that central product more appealing.

LEGO also realized that it couldn't just sell its signature product—the plastic building set—without surrounding it with other little innovations such as comic books, video games, movies, or events at LEGO stores. It changed its product development process to continually develop a full portfolio of complementary innovations around every major new toy line.

A FAMILY OF INNOVATIONS FOCUSED ON FULFILLING A SINGLE PROMISE. The family of innovations must be aligned around a central theme or strategy that we call the *promise*. Again, this is precisely what Apple and LEGO did. Apple's music system was a key part of its overall promise to users in 2001: we will help you manage your digital life, starting with your music. What produced Apple's success was not a haphazard collection of insanely great innovations but the way those innovations created a system of interdependent features and capabilities that worked together to fulfill the Apple promise. None of Apple's key products could have succeeded on its own. Imagine the iPod without iTunes in 2001: Would its impact have been anywhere near as large?

LEGO also focused on its promise of delivering a compelling story-based play experience. Every book, movie, game, and event was carefully crafted to involve kids in the drama LEGO created around each toy line, making those boxes of plastic pieces irresistible.[25]

A FAMILY OF CENTRALLY MANAGED INNOVATIONS. The careful selection and proactive management of a family of complementary innovations is both the essence of the Third Way and what makes the approach so difficult to execute. Consider the challenge faced by Apple Computer in 2000. To deliver on its promise, Apple needed to develop a new music player, software for managing music, and a network of retail stores to sell it all.

When Apple was looking for software to manage MP3 music, it looked outside the company and acquired the best competitor in the market, a program called SoundJam, which had been created by former Apple engineers. Apple acquired the rights to SoundJam in 2000 and reintroduced the program as iTunes in 2001. To develop a first prototype of the new player, Apple hired most of the development team from outside Apple.[26] It sourced the chipset from a San Jose company called PortalPlayer, earbuds from a company called Fostex, and the operating system from Pixo, a Cupertino company. At every step, Apple went outside the company to find the solutions and expertise it needed, but retained close control over all the complementary innovations.

Similarly, LEGO has also kept tight control over the products and services that help deliver on its promise. It used its chain of LEGO stores not only as retail points of sale for its products but also as venues for events to promote its new toys and involve kids in the stories around them. When LEGO was looking for PC games to complement its toy lines, it formed a close partnership with TT Games, a company headed by two former LEGO executives who had been laid off from the company in 2003. When the company decided to invest in *The LEGO Movie*, it partnered with the team that had created the movie *Cloudy with a Chance of Meatballs*, but it made sure to retain creative control over the story and characters in the movie.*

The Companies Followed a Similar Process

A second reason we discuss the Apple and LEGO turnaround stories together is to show the remarkable similarities in the process they both followed. The similarity is probably not a coincidence; when LEGO was searching for ways to save the company in 2003, it could

*Our favorite example of this control is the policy that the male and female LEGO minifigures in *The LEGO Movie* could hold hands, but not kiss.

use Apple's turnaround, which had begun just two years earlier, as a model. The process each followed can be summarized as a sequence of four decisions.

DECISION 1: WHAT IS YOUR KEY PRODUCT? Before deciding where to innovate, Apple and LEGO first had to decide what product to innovate around. For LEGO, the decision was fairly easy—loyal customers and retailing partners by 2003 were telling LEGO in no uncertain terms that if it didn't return to the brick, then the company was unlikely to survive. For Apple, its core or key product was the Mac, not—as many mistakenly think—the iPod or iTunes. The music player and software were designed not as new business opportunities but as complements to the Mac—as products and services designed to make the Mac more desirable.

DECISION 2: WHAT IS YOUR BUSINESS PROMISE? Both Apple and LEGO next developed a compelling business promise—a commitment to the customer that would help make the key product more desirable and valuable because it satisfied some compelling customer need. LEGO promised to make the play experience around a toy line richer and more exciting. Apple promised to help customers manage their digital lives.

DECISION 3: HOW WILL YOU INNOVATE AROUND YOUR KEY PRODUCT? After a company chooses a promise, it has to choose the specific innovations that will work with the key product to deliver on that promise. Apple not only developed the iPod and acquired iTunes, but also launched its line of Apple Stores in 2001 as a way to communicate to the customer the benefits of its hardware and software offerings. LEGO works closely with kids to discover and develop ways it could enrich their play experience around a plastic toy line.

DECISION 4: HOW WILL YOU DELIVER YOUR INNOVATIONS? Every step along the way, both Apple and LEGO made acquisitions, developed partnerships, or hired internal talent in order to retain tight control over all the different products and services it launched.

We will explore these four decisions in greater detail in the chapters to follow, where we will also identify the managerial and organizational problems that each decision can create. We hope the Apple and LEGO stories illustrate the potential rewards from executing the Third Way well.

Sometimes, Little Ideas Become Big Ideas

Apple's experience also illuminates how key products and the complementary innovations around them can evolve. In 2001, Apple Computer's core product was the Mac and all the other innovations, including the iPod itself, were complements meant to lure Windows users into the Mac world. But when Jobs saw the iPod's popularity, particularly after its redesign in 2007, he made it part of the company's core offering, which would come to comprise not just the Mac but a small stable of other key products—the iPod, iPhone, and iPad—that singly or together allowed users to better manage their digital lives.

This change became obvious when Apple released iTunes for Windows. Apple realized that this family of innovations could not only help draw people to the Mac, but also could generate profits for the company even if deployed around a competitor's PC. What started as a way to generate interest in Apple's key product—the Mac—became a standalone set of products that now generate seven times more sales than Apple's PCs do.[27]

Similarly at LEGO, the development of small innovations such as stories and games around each product line led to the big innovation of a full-length animated feature film in 2014, *The LEGO Movie*. While developing an animated feature film is by no means a new-to-the-world innovation, LEGO's move into this market has the potential to open up opportunities in the much larger entertainment field. And the expertise it acquired in making the computer-generated movie, in which every element looks like it was made out of LEGO bricks, will make the next movie much easier to develop.

The LEGO and Apple stories illuminate the potential of the Third Way to unleash vast change and build strong competitive positions—positions that competitors struggle to match because they're a complex, interdependent mix of multiple and diverse innovations. In pursuing the Third Way, both Jobs and Knudstorp demonstrated that you needn't disrupt yourself to be disruptive or to carry out major competitive strategies. Neither Jobs nor Knudstorp put their companies or brands in jeopardy. And both illustrated how building around a core product can make that product more desirable, more valuable to customers, and more profitable.

We hope that these two stories illustrate how powerful the Third Way can be for extracting value from a product. Now the question is how to put the approach to work in your organization. For most organizations, it can present difficult organizational and managerial challenges. In the next chapter, we will expand the basic four-decision framework we've outlined here and explore the source and nature of challenges raised in each decision and how they can be addressed.

Three Takeaways for Chapter 2

- Both LEGO and Apple Computer achieved dramatic company turnarounds without disrupting their core product. Both had tried and failed to enter new markets with new types of products. To restart growth, both developed or acquired a set of low-risk, complementary innovations around their core products—the brick and the PC—that greatly enhanced the value of those products.
- Between 1997 and 2007, Apple Computer succeeded because it was a master at developing a family of complementary innovations that helped customers get more value from its core product—the PC. While Apple later introduced such revolutionary products as the iPhone and the iPad, those products were successful in part because they built on the foundation of complementary innovations that Apple Computer developed between 1997 and 2007.
- For Apple and LEGO, the Third Way was the only way for each company to survive and thrive in tough, competitive markets.

3

The Four Decisions and Why They're Difficult

CarMax, the used-car superstore we discussed in chapter 1, was an almost immediate success. After launching its first store in 1993 and working the bugs out of its systems, the company launched several more over the next few years. In 1996, it announced its plan to open fifteen to twenty new stores per year. This early success quickly attracted serious competition, including AutoNation, Car America, Driver's Mart, and Car Choice.

By far the most dangerous of this lot was AutoNation, a company led by Wayne Huizenga, the billionaire entrepreneur and Wall Street hero who had built two wildly successful *Fortune* 500 companies from nothing. Starting with one garbage truck in 1968, he grew Waste Management International into the country's largest waste disposal company through rapid acquisition of local firms. Then, using the same process in the 1980s and early 1990s, he built Blockbuster Video into the nation's largest movie rental chain. From there he went on to buy the Miami Dolphins football team, the Florida Marlins (who would win the World Series in 1997), and the Florida Panthers (hockey).

Setting his sights on the used-car market, Huizenga acquired AutoNation and Car Choice and began rapidly opening used-car superstores in major cities across the Sunbelt, precisely the region where CarMax was planning to concentrate its efforts. Yet by the end of 1999, just three years after entering the market against CarMax, AutoNation announced that it was shutting all its used-car superstores and exiting the business. One of the greatest entrepreneurs in US history had been unable to beat a group of "washing-machine salesmen from Richmond, Virginia."[1]

The AutoNation story illustrates the challenges of executing the Third Way. Like much else in management and leadership, the underlying concept of the Third Way is simple and straightforward, but its implementation is difficult. The root of the difficulty resides in its first and most basic feature—that it's a diverse set of complementary innovations around a key product. The need for diversity moves the innovation challenge beyond product groups, where it normally resides, to business functions—manufacturing, legal, finance, procurement, and so on—that typically aren't asked to innovate. In many companies, they're not even asked to collaborate.

Many functions have always been involved in innovation, albeit in other ways. Legal, for example, is routinely asked to vet a new product for potential legal problems. But with the Third Way, lawyers might also be asked to search actively for an innovative legal approach that makes a product more attractive, to set up contracts with outside partners to fund their product development efforts, or to assess the legal risk of some new business model. Similar challenges will stretch the capabilities of procurement, manufacturing, marketing, finance, and business development.

Involving such a wide variety of groups raises problems, ambiguities, conflicts, and questions that many organizations find troublesome, especially those that prefer the neat separation of roles and functions.

Innovation calls for creativity, an inherently messy process. Because most organizations strive for clarity and order, groups in them may struggle to do the kind of creative cross-silo collaboration and innovation that the Third Way requires.

We don't focus on these challenges to discourage you. Our aim instead is to prepare you for the hurdles ahead. Indeed, the remainder of the book will be dedicated to precisely that goal. As you deal with these problems, remember that the difficulties of successfully pursuing the Third Way are what make it such a powerful competitive weapon.

To see how these challenges literally determined the fate of one company, let's return to the story of CarMax and its most dangerous competitor, AutoNation.

Why AutoNation Couldn't Execute

The first CarMax store opened in Richmond, Virginia, in 1993 and exceeded all sales and profit targets. The second store—opened the following year in Raleigh, North Carolina—was also a success, and the company prepared to move ahead with a chain of superstores. CarMax continued through that year and 1995 to perfect its processes and systems, especially the custom IT system that would be the backbone of its operations.

Both the Richmond and Raleigh stores had offered a selection of approximately four hundred cars. Knowing how important selection was to customers, CarMax wondered if more cars might be even better. So it decided that one of its next sites would be a thousand-car megastore serving a major metropolitan area. For that purpose, it chose Norcross, Georgia, a suburb of Atlanta, and opened a megastore there to great fanfare in 1995. Later that year, the company opened a second megastore in Maryland. Both surpassed all expectations, and so the company proceeded to roll out a combination of small, medium, and large location sizes, depending on the population of the region served.

Norcross, the first megastore, was a turning point for CarMax in another way as well. Until then, the company had purposely worked quietly in smaller markets to avoid drawing attention to itself. It wanted time to perfect its approach and build momentum. But Norcross, a major test in a major market that was necessarily undertaken with much fanfare and publicity, did not go unnoticed. If CarMax had initially found for itself a blue ocean market with little serious competition, the water after Norcross wasn't going to stay blue for long. The sharks now smelled blood and rushed to attack.[2]

Several copycat startups—AutoNation, Car America, Driver's Mart, and Car Choice—quickly jumped into the market. The most aggressive was AutoNation, which used the reputation of its founder and the funds he had generated from his other businesses to expand rapidly across the Sunbelt by snapping up the best locations in all the most important cities.

AutoNation upped the bet in other ways too. It began selling new cars as well as used cars to improve profitability. To speed growth, it began acquiring existing dealerships and incorporating them into the AutoNation chain. And it began to acquire the other major chains that had sprung up in response to CarMax's initial success. Eventually it acquired all of them, turning the struggle into a two-company showdown for supremacy and survival.

CarMax had to respond. It generated funds through an IPO in 1997 and initiated what its cofounder, Austin Ligon, called "the shoot-out across the Sunbelt." To keep up with AutoNation, it had to expand much faster than it had originally planned. Competing against a proven company builder, CarMax struggled to keep up. After two years, in 1999, it seemed to be losing the shoot-out. AutoNation had opened forty used-car superstores while during the same period CarMax had managed only thirty. CarMax's stock slowly sank below $10 in 1997, less than half its IPO price, and kept sinking to a low of around $2.50 in November 1999.

Yet while CarMax was struggling, AutoNation's problems were far worse. Its new-car operations were profitable, but not its used-car business. Losses there in late 1999 were running an estimated $25 million a month. Huizenga brought in new leadership, which looked hard at the $500 million in used-car losses and investments to date. On December 13, 1999, AutoNation closed every one of its used-car superstores, instantly returning the used-car market to the pristine state CarMax had found in 1993.

Why did CarMax win? It certainly wasn't perfect execution. Its decision to build thousand-car superstores after the success of its Norcross store was an almost fatal mistake.[3] But it never abandoned its original approach, which was to offer a great selection of used cars and make the car-buying experience pleasant and stress-free for customers. This was where AutoNation failed. It had been able to match CarMax's selection—it even sold new cars too—but as much as it wanted to emulate what CarMax was doing, it couldn't deliver the stress-free experience.

AutoNation's strategy of growing rapidly through acquisitions, rather than building each store from scratch as CarMax did, meant it inevitably brought onboard existing systems and personnel from acquired dealers, with all the high-pressure sales tactics that car buyers hated. AutoNation also lacked the underlying organizational infrastructure—especially the standardized IT systems—that provided data-driven insights and, most important, enforced the practices that made the experience trustworthy and stress-free, such as no-haggle prices.

AutoNation chose to make many of its outlets full-service dealers that offered not only used cars but also new cars as well as auto service and repair. A broader offering might seem to offer real advantages—after all, AutoNation was making money on new cars—but the additional complexity prevented it from focusing its entire organization on the practices that the used-car business required for success.

In short, no matter how much it wanted and intended to match the experience provided by CarMax, AutoNation simply lacked the necessary organizational skills. It couldn't execute.

A key reason CarMax could execute was that it took the time in its early years to experiment with different ways of providing a better buying experience and then to build the systems and processes that supported and enforced consistent execution of the lessons learned. Its early IT investments were costly, but the costs decreased once the underlying systems were completed. For AutoNation, rapid growth seemed to trump all other priorities.

When AutoNation shuttered its used-car business, CarMax halted all growth for two years to refocus and correct deficiencies, such as the megastores, that had crept in during its headlong race with AutoNation. Thereafter, it resumed growing, with sales of $6.3 billion and profits of $134 million in 2006, expanding to $14.3 billion in sales and $597 million in profits in 2015.

CarMax's success contained more than a little poetic justice. When AutoNation closed its used-car business, CarMax's stock sank even lower, from $2.50 on December 14, 1999 (the day after AutoNation announced its exit), to less than $1.50 in early 2000. According to the wisdom on Wall Street, if the great Wayne Huizenga couldn't make the used-car business work, no one could—certainly not those "washing-machine salesmen" from Richmond. But Austin Ligon and his colleagues considered the low stock price a huge opportunity. Confident in the approach they had laid out in the beginning, they exercised their options and bought more stock. And when CarMax succeeded, they made personal fortunes.

The story of AutoNation is a concrete illustration of the point we made earlier: though simple in concept, the Third Way poses management challenges that make it hard to pull off, even for a great entrepreneur. It spreads the innovation challenge across all or most groups in the company. That means those groups, plus outside partners if needed, must work together in new ways. Groups with different ways of seeing, thinking, and talking, as well as different performance goals, metrics, and systems, must create innovations that operate seamlessly for the customer. That's what CarMax was able to do but AutoNation could not.*

The Four Decisions

In chapter 2, we briefly introduced the four decisions, a simple process for pursuing the Third Way that was drawn from the turnaround stories of Apple and LEGO. As described there, these four choices are a good way to explain the "how" of this approach to innovation, the key phases of activity needed to pursue the Third Way successfully. They're also a good way to understand where, why, and how the challenges of the Third Way are likely to arise and why CarMax prevailed over AutoNation.

FIGURE 3-1

The four decisions

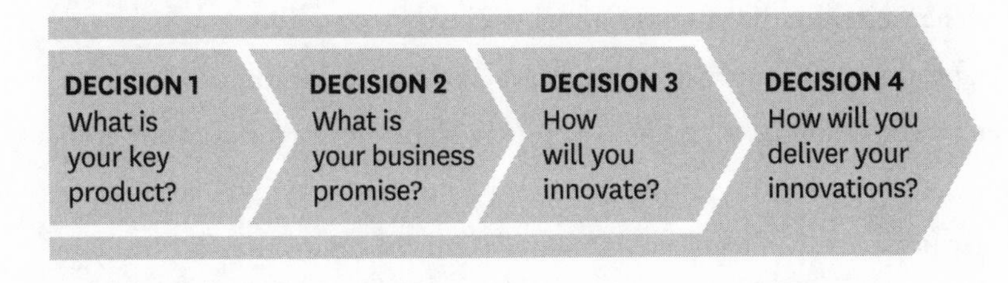

*The CarMax story illustrates a second point, one we made earlier: the Third Way is not an innovation approach mutually exclusive with other approaches—CarMax used the Third Way to support its blue ocean offering. AutoNation tried to implement a similar strategy, but didn't surround its core product with a Third Way system and failed as a result.

Decision 1: What Is Your Key Product?

For most organizations and in most settings, choosing a key product is a straightforward decision: what is the product you wish to make more attractive to customers? For most of the companies we've discussed, the key product was also the company's core product, its heart and soul. But as we saw with Novo Nordisk and Norditropin, it need not be. A key product can be any product that meets two basic requirements.

First, it should be one of your crown jewels.[4] Ask yourself, will selling more of it help our firm achieve its strategic goals? Will increasing the sales of this product reflect and advance who we are in the marketplace? Is it a product that customers associate with our company? Applying this approach to a product that's not one of your crown jewels won't be worth the effort. Even if you manage to increase sales, it won't take your company where it wants to go.

Second, a key product must be a product that's fairly stable; that is, unlikely to change significantly in the medium term—the next, say, three to five years. Is it something you offered yesterday, offer today, and will be offering tomorrow? It's difficult to build a system of complementary innovations around a product that will soon undergo more than minor alterations.

When CarMax began operations, it clearly focused on one, and only one, product: the nearly new, used car. CarMax's focus on used cars was seen by the financial markets as a constraint on the fortunes of the company. AutoNation, by selling both new and used cars, was seen as the better prospect. In fact, CarMax's focus had the opposite effect: only by focusing on one product was the company able to surround that product with the complementary innovations it needed to succeed.

If your organization has several products that qualify, you'll need to determine which few offer the greatest opportunity. The Third Way requires enough effort from many groups that choosing more than a few key products at a time could be difficult.

Decision 2: What Is Your Business Promise for Your Key Product?

This is the heart of the Third Way. The promise is the commitment that ties together all complementary innovations around a key product. It is a statement that commits the organization to deliver more than just the product; it will help customers apply or use the product to accomplish their goals. In virtually all cases, a promise is based on the customer need or desire that the key product with its surrounding innovations will satisfy. For LEGO, it was the promise of a rich, immersive story that a child could experience in a variety of ways. For CarMax, it was the promise of a pleasant, stress-free used-car-buying experience, a promise that AutoNation wanted to duplicate but couldn't. For Novo Nordisk, it was the promise of a less painful, more convenient, and easier-to-remember HGH therapy process.

The promise, in other words, is a commitment to solve a pressing problem or fulfill an unmet need for the customer. It says the key product and complementary innovations around it will provide a complete solution. Because, ideally, it will be a solution not provided by competitors, the promise is what distinguishes your product and sets it apart.

CarMax's promise—a hassle-free, trustworthy used-car-buying experience—was difficult to deliver, and it was even more difficult to convince customers that the company would deliver that experience. CarMax salespeople said customers entering the store would give them "the Heisman," a reference to a US football trophy that portrays a player with one stiff arm extended to ward off opponents. CarMax's goal was to convince skeptical customers that it was different. Everything it did, from the design of the products to the layout of the stores, was created to deliver on this promise.

AutoNation tried to emulate some aspects of that promise, such as the no-haggle pricing, a large selection of vehicles, and a full suite of other products like loans and insurance. But its strategy of expanding through acquisition forced it to use existing IT systems, sales processes, and (most crucially) employees—in short, the old way of selling used cars. As a result, it was unable to authentically deliver a better experience to customers.

Decision 3: How Will You Innovate Around Your Key Product?

Here you identify the specific, tangible innovations that fulfill the promise you've chosen. First, you identify multiple candidates. Then you test them through experimentation, prototyping, and research. Finally, you select those that truly deliver on the promise. When CarMax realized that the opportunity was to provide a stress-free buying experience, it then had to take apart the entire buying experience, identify the exact sources of pain and annoyance—haggling over price, for example—and then test ways to remove the stress from each step. Along the way, it made many mistakes, but by diligently focusing on delivering its promise, the company ultimately drove all its competitors out of the market.

Decision 4: How Will You Deliver Your Innovations?

Finally, you must either locate or develop the specific innovations you've chosen. The first step is to identify who will be responsible for each innovation. These leaders will be people spread around the organization, depending on the nature of each innovation. And they're likely to include outsiders who possess specific, unique skills—IT developers, for example—that your organization lacks. The second step is to decide how you will manage these diverse people and groups, that is, what organizational roles and practices will be necessary. This second step must begin very early in the process to ensure that the project leader has the skills and resources on the team that are needed to manage each step of the process.

What Makes Each Decision Difficult?

The four decisions seem straightforward and even simple, at least in concept. Why would they create problems?

Decision 1 Challenges: What Is Your Key Product?

Recall what Steve Jobs did with Apple's product line when he returned in 1997. He conducted a thorough review in which product people, one group after another, presented their lines to him and explained why they were important to the company. Then he killed 70 percent of the products, eliminating almost everything except four Mac models.

How do you suppose the product people associated with the 70 percent reacted? Probably not happily. Of course, that was an unusual situation, where products and models had proliferated like rabbits for years and drastic action was appropriate. But the point remains: to pick one or a few key products means that you're deciding that the other products are not key. Consequently, everything else has to be eliminated, turned into a complement, or deemphasized. Rationalizing a product line is bound to create what seem to be winners and losers. Positions will disappear or be downgraded, and the incumbents moved to new positions or terminated. It's hardly a happy process for any company, but could Apple have succeeded if Jobs had retained all or many of the 1997 product lines? That's hard to imagine.

The second potential problem arises from the condition of many products that might be key product candidates. It's known as the *coral reef syndrome*, based on what happens to large structures dropped into tropical oceans. Say you dropped a new luxury car into an ocean—what would happen? It would still look like a car for awhile, but after a few years, it would become inhabited with all forms of sea life.

Over time, these small sea creatures would live and die, leaving their shells behind, and that once-beautiful luxury car, now encrusted beyond recognition by marine plants and animals, would look like a coral reef. A similar process occurs with products after they're launched. So many changes, revisions, and "improvements" are made that the once-beautiful product turns into something unrecognizable.[5]

Obviously, a product whose function and benefits have become blurred and confused cannot function well as a key product. So many features may have been added to broaden the product's appeal that its target audience is no longer clear. Before it can serve as a key product, it must be rehabilitated, and that requires time, effort, and other scarce resources. Whatever features and product line extensions have been added, someone in the company is likely to have a stake in them and will resist change, not to mention the inevitable loss of customers, however many or few, who were attracted to each of those additions.

In short, a product chosen as a key product must be made as simple, lean, and strong as possible. Otherwise, there will be confusion about the promise it fulfills and, consequently, confusion about the complementary innovations to be added around it. Note that "simple, lean, and strong" refers to more than the product itself. Jobs outsourced manufacturing to reduce costs, as did LEGO once it discovered the Third Way.

If a key product requires more than incremental improvement, you must make those more serious changes before embarking on the Third Way. It's dangerous to create a system of complementary innovations around such a key product because, as you change it, you probably will have to go back and change (or even drop) complementary innovations that are no longer compatible.

Finally, the need for a key product to remain stable after installing complementary innovations around it can create problems too. Conventional wisdom today says nothing should be off-limits to change, even to radical change. To decree that a key product cannot be changed significantly will require courage and discipline, especially if the Third Way has been chosen as the response to a serious competitive threat.

Decision 2 Challenges: What Is Your Business Promise for Your Key Product?

The ideal promise—the benefit a key product or service provides the buyer—possesses these basic characteristics:

- It's based on a need or desire that's important to the customer—such as LEGO's promise to give kids a rich, compelling story that they can play out with a construction toy and the many complementary innovations (e.g., a movie) around it.
- It fills a need or desire not satisfied well by competitors—for example, the trustworthy buying experience only CarMax was able to provide.
- It draws on the particular expertise of the organization—for example, Apple's unique ability to meld hardware and software in ways that create a better user experience.
- It creates, extends, or reinforces the public perception, that is, the brand, of the company—for example, Novo Nordisk's ability to provide superior products for patients who must inject themselves daily.

The strength of a promise, namely, its ability to attract buyers, will depend most directly on the depth and urgency of the customer need it fulfills, especially in comparison with the way competitors fill (or don't fill) the same need. Consequently, choosing a powerful promise will depend on your deep understanding of customers and their needs.

This understanding seldom comes from focus groups, market research, or any other form of reductive market analysis.

It comes instead from watching and learning from customers—putting yourself in their lives and understanding the context in which they choose and use your product. Though data and analysis are certainly useful, finding a powerful promise depends even more on the ability to empathize with customers, which is as much a creative task as it is analytical. This kind of insightful understanding is hard to develop. Often it requires the kind of touchy-feely research that in some organizations is hard to justify because it doesn't produce clear, definitive answers.

It's highly unlikely that data or analysis will ever prove empirically that a specific promise is the right one. CarMax couldn't prove that providing a stress-free buying experience would attract large numbers of customers. The data, such as it was, only provided a hint, a direction, as to where in the experience of buying a used car an answer—that is, a powerful promise—might be found. LEGO had to try all manner of approaches with new products before one of them worked and people at the company could deconstruct why.

Because they indicate a direction, an area, where a strong customer need is likely to be found, promises are necessarily conceptual rather than concrete. "A stress-free car-buying experience" is an idea that exists one rung on the abstraction ladder above a specific, concrete innovation—such as no-haggle prices. Because it's conceptual rather than tangible, a promise is likely to be dismissed or ignored by those who prefer going straight to "what to do."

That disdain will be warranted if the promise is too abstract, if it's so vague that it's meaningless. Suppose CarMax promised to provide a better car-buying experience. This kind of generality is neither appealing nor useful, whereas the promise of a no-hassle, trustworthy car-buying experience, while still conceptual, is far more useful. It says, Find the sources of customer stress and discomfort in the buying experience, and remove or neutralize them. That's what CarMax did—with great success. So not only is a good promise grounded in a real and compelling customer need or desire, but it's also expressed in a way that suggests specific innovations that can be tested.

Finding a good promise is a step some companies don't bother to take. They go straight from choosing a key product to choosing the specific innovations they will add around it. The danger of skipping the promise is that the innovations then added will never be more than a random collection of unconnected good ideas that don't work together. Done right, a strong promise will produce a family of innovations that collectively are much more powerful than the simple sum of the pieces.

Decision 3 Challenges: How Will You Innovate Around Your Key Product?

This is the stage where you identify, test, and select the actual complementary innovations that fulfill the promise. Like so much else about the Third Way, it seems perfectly straightforward when, in fact, it presents a host of traps that can cripple the whole approach.

Needless to say, these steps, because they're innovations, are by definition new and unfamiliar, with all the problems that can cause. But that's only the beginning of what can go wrong.

The Third Way calls for a diversity of innovations that cut across many business functions. This is a strength of this approach. Diversity makes it harder for competitors to duplicate the approach, and it gives the company flexibility in where it makes money. But this approach means working across silos to find and engage people who've probably never done this kind of innovation before. Will they be interested? Will they even care? Will they be allowed to devote the time necessary? Above all, will they be able to come up with anything that actually makes the key product more attractive?

Then there's the problem of selecting the innovations to pursue from all the candidates identified. This step is especially hazardous because the evidence is clear: managers, even experts, are poor at choosing winners (a topic we'll discuss more in chapter 6). Success requires an openness to learning from experience, data, and constant experimentation. Too often, leaders consider it a key part of their role to make these choices based on instinct, gut feel, and personal insight. Alas, this approach—often called the HIPPO method because the decision is based on the highest-paid person's opinion—produces more losers than winners.[6] Like the promise, these choices are made best when they're based on a deep, data-driven, and experience-driven understanding of the customer and the context within which he or she uses the key product.

You needn't develop and launch all complementary innovations at once, but as we mentioned earlier, there will be a minimum viable portfolio of complementary innovations—a kind of critical mass—that must be in place to satisfy the basic requirements of the promise to customers. Think of the iPod when it was first released with only the original iTunes to support it. There was interest but nothing like the avid response generated by the launch of the iTunes Music Store and the subsequent expansion into books, videos, and apps.[7]

For your key product and surrounding innovations, you will need to understand this minimum set and develop it as quickly as possible. As explained, this decision needs to be based on data and customer knowledge, not the HIPPO. Once you have this minimum set of innovations in place, you can and should continue to improve and add to it through ongoing research and testing.

As you select complementary innovations, you will inevitably come across some that seem attractive, even too good to pass up, except for one small problem: they don't support the promise.

AutoNation couldn't resist the urge to grow through acquisition and to expand into new-car sales. This move certainly turbocharged its expansion and solved its procurement challenges. But it also prevented the central office from enforcing a no-haggle, trustworthy sales process, ultimately allowing CarMax to prevail. In spite of everyone's best intentions, these quick-profit opportunities will be tempting. Some will argue for "just this one exception." You will need discipline to stay focused on the promise.

A convenient way to do this is to remember that you must be able to constrain, connect, and control all complementary innovations:

- *Constrain:* Make sure each innovation helps fulfill the guiding promise and aligns with the brand.
- *Connect:* Make sure each innovation fits with the other innovations and the key product to create a coherent team or family, not a random collection.
- *Control:* Make sure you can manage each innovation: who does it, its constraints (e.g., budget) and standards (e.g., quality), the definition of success, and the governance process (i.e., who makes what decisions).

Finally, some complementary innovations may require the use of outside partners that possess crucial skills your organization lacks. For many organizations, this necessity is likely to raise unfamiliar issues and challenges. How will you ensure that an innovation developed outside will be constrained, connected, and controlled? Making something new obviously involves the creative process, which is usually nonlinear and often untidy. For many organizations, this is unexplored territory, and they will need to install new management practices and processes that provide necessary control without stifling the innovation.

Decision 4 Challenges: How Will You Deliver Your Innovations?

You've chosen the key product, identified the promise to be fulfilled, tested and selected the specific innovations to pursue, and now, finally, you must identify who will actually do each innovation and how you will manage them. Again, what sounds like a straightforward management task can frequently raise crucial, make-or-break challenges.

As we've seen, the fundamental source of these difficulties is the need to involve groups that have never innovated or otherwise worked together in this way before—groups that by training, inclination, and experience speak different languages, seek different goals, and measure themselves by different standards. To make this approach work, most organizations find they must install and practice new ways of managing.

We saw something of this in the story of CarMax, which as a startup worked hard and took the time necessary to install and perfect the systems, practices, and even physical structures needed to deliver on its promise.

Imagine calling a meeting of people from business development, marketing, and finance, all of whom have their own jargon that reflects different ways of seeing the world, and then telling them, "You're going to create complementary innovations in your areas that all work together seamlessly to support this key product. Also, some innovations will be done by outside partners whom you'll have to supervise. And every innovation must be ready to go in six months." That would be a hard challenge, but then add to it the complication that you will ask many of these groups to forgo profits, take on additional costs, or even lose money for the good of the whole.

Every complementary innovation comes with costs that must be borne by the group creating and managing it. Those at Apple in charge of the iTunes music store, for example, had to accept that music sold through the store would be sold basically at cost. From the beginning,

Apple believed that music revenue should go largely to the music companies because that was best for Apple overall. Some complementary innovations should forgo possible revenue because it's part of a larger effort. Sales managers at CarMax had to accept no-haggle prices because the company believed customers attracted by that feature and by the additional services they would buy would more than compensate for any lost revenue that haggling might have produced.

Similarly, LEGO had to reorganize in ways that let it use inside groups or outside partners to create a multitude of stories and games centered around each of its construction toys. As a startup, CarMax had the luxury of organizing from the beginning around its basic promise. AutoNation couldn't match that promise because, from the beginning, it organized itself as a traditional dealer. To change over to CarMax's approach would have required AutoNation to make wholesale changes in its personnel, practices, culture, processes, and systems—obviously a huge challenge and one it didn't meet.

Using outside partners brings even more complexities and problems. When you work with inside partners, at least all of you in theory work for the good of the same organization. But with outside partners, there will inevitably be different motivations. How do you work with outsiders so that their innovations are constrained to satisfy the promise, connected to work seamlessly with each other, and always in your control? Most organizations haven't done this before and don't know how to do it.

It doesn't help in all this, of course, that as an insider, you by definition don't know much about what such outsiders do and the markets they operate in. How did LEGO, which knew virtually nothing about making movies, work with the producers and director of *The LEGO Movie*? When the company in 2001 hired outside producers to create a TV series based on its Galidor toy (a buildable action figure that came out at the same time as Bionicle), the result was an embarrassing disaster, a terrible show that hurt sales of the toy.

Dealing with these many problems will challenge virtually any organization, especially one pursuing the Third Way for the first time. The reality is that while the Third Way makes customers' lives easier, it almost invariably makes organizational life more complex and difficult because it turns people in virtually all functions into innovators. The more structured and traditional an organization is, the more trouble it will have. That's why this decision—how to deliver the innovation—raises probably the most perplexing challenges of all.

What Do the Four Decisions Require?

As we've just seen, each decision presents its own set of difficulties, and these challenges will be the subject of the remainder of this book. We can, however, conclude this chapter with three pieces of advice:

To Manage a Third Way Project, Use a Third Way Process

Hundreds of books have been written about how to structure an innovation process, and just as innovation is often defined as being either incremental or radical, the recommendations for structuring an innovation process tend to fall into two camps: some argue for systematic, disciplined methods, and others for an iterative, experimental approach. Advocates of disciplined processes argue for structured techniques, regular reviews, and specific deliverables.[8] They believe that complex innovation projects need careful support and rigorous management.

Advocates of iterative experimental approaches argue that the uncertainty of an innovation project requires flexibility in the process.[9] An iterative experimental approach has rapid build-test cycles, incremental learning, and (hopefully) steady progress toward a successful result. Structured research and development methods are still important parts of the process, but the way those methods are chosen and combined at each phase depends on the results of the previous phases.

A Third Way project requires a third type of process. Structure and discipline are required across the entire process to ensure that the central role of the key product is maintained and the important elements of the promise are delivered. But the range of different possible complements and the uncertainty around their delivery will require experimentation and iteration throughout the process. While the four decisions may appear to be a sequence of four steps to be done in order, at times you will find yourself moving back and forth among them. You may find that what you learn in the second, third, or fourth decisions will lead you to revisit previous decisions and modify them.

For example, work on decision 2—identifying the promise of a key product—might provide insights that cause you to go back to decision 1 and rethink what is and is not a key product. When CarMax defined its promise as a trustworthy, hassle-free car-buying experience, it realized that this promise could only be delivered if the used cars were fairly new. Older cars were inherently less reliable, but more importantly, selling older cars would have led to a variable commission structure. CarMax realized that to ensure a great car-buying experience, its salespeople had to be motivated to sell but not to oversell, so it decided to put a fixed commission on every car. No matter which car a customer chose, the CarMax salesperson got the same commission. This arrangement, coupled with CarMax's no-questions-asked, five-day money-back guarantee, ensured that every salesperson was motivated to sell only the car a customer really wanted, nothing more and nothing less. If CarMax had expanded into older car sales, it would have expanded its potential market, but hurt its ability to deliver on its promise.

When you are making the second decision, difficulty in finding a compelling promise for a key product might also lead you to rethink whether that product truly is a key product. In the same way, problems with decision 3 (identifying specific innovations for satisfying a promise) could lead you to conclude that the promise cannot be fulfilled, which, of course, might even lead to rethinking the key product itself.

There's nothing inherently wrong with this kind of looping back to previous decisions. Indeed, you should always be willing and ready to do it, because it simply reflects the reality that the whole process is inherently creative and thus will often be messy. It's a fine balance to maintain. On the one hand, innovating can be a little chaotic; on the other, clarity and discipline in each step are needed to ensure that the later steps support the key product and deliver on the promise.

Involve More People Across the Firm in Innovation

Here we emphasize a point already made: that innovating across multiple business functions will require the involvement of many people drawn from those diverse groups. No longer can innovation be handled solely by people from the product development and product management groups.

Product developers have never worked entirely alone. They've always had to coordinate with manufacturing, legal, finance, marketing, procurement, customer service, and other groups that would be involved in making, selling, and supporting a new product. But that kind of involvement is not what we're talking about here.

The Third Way calls for something more, a new kind of involvement for many of these groups. They are being asked to become innovators and innovation managers themselves, to create and manage new ways of manufacturing, selling, fulfilling, financing, supporting, advertising, or whatever their groups do, in order to make a key product more attractive to buyers. In other words, they are now being asked to create specific complementary innovations, related to but separate from a key product, that help fulfill the promise of the product. That's not the role they've typically played, and it's a role that requires a significantly different set of skills.

This is where organizational problems can arise because in this new innovation-related role, these disparate groups will need to be managed not just by their normal superiors but also by the leaders who are managing the Third Way for their products. It may sound as if this situation calls for a cross-functional team, a device many organizations already use frequently. But this is not that.

The Third Way calls for an ongoing commitment of time and resources from the innovating group—finance, for example—and those resources are likely to be scarce. Finance leaders may prefer not to expend them this way, and they may not want their people being managed by people in some other part of the organization. Sorting all this out and keeping it running smoothing will fall on the product manager, for whom the responsibility will probably be new and daunting.

In short, the Third Way is likely to change any organization pursuing it because far more people will be involved in the act of innovation—not just supporting the innovators but being real product-related innovators themselves. This new level of involvement represents a sea change in most organizations.

Redefine the Product Manager Role

A key consequence of the Third Way is that it forces the product manager out of his or her traditional role. Simply put, it changes that leadership role from directly managing product development to managing the four decisions, which means managing the diverse set of people inside and outside the organization to deliver a full portfolio of complementary innovations. That's a significant change, to say the least.

In this new approach, product managers are still necessary, as before, to shepherd new products through the development process. But there needs to be a higher-level role, a *solution integrator*, who drives the cross-organizational process for designating a key product; finding the strongest possible promise for that product; selecting, specifying, and designing a complete set of complementary innovations; and choosing the partners from inside and outside the company to deliver those innovations. To do this, the product manager will need to build and manage a cross-functional team and report to a supervisory team whose members have the necessary breadth of authority.

As noted earlier, this chapter has focused on the problems of implementing the Third Way, not to discourage you but just the opposite: to keep you from being discouraged when problems inevitably arise.

The four decisions and the challenges unique to each are the focal points for the remaining chapters. In each, we will work through one of the decisions and reveal the insights, guidelines, best practices, and working principles needed to pursue it successfully. By the end, you and your organization will be prepared to undertake this low-risk, high-reward third option for dealing with a perilous, fast-changing world.

Three Takeaways for Chapter 3

- Four decisions summarize the key choices that make up the Third Way: (1) What is your key product? (2) What is your business promise for your key product? (3) How will you innovate around your key product? (4) How will you deliver your innovations?

- The goal of these four decisions is to bring to market a family of innovations that will deliver on the promise. Doing so will challenge the Third Way team in new and unfamiliar ways, and will require changes in the innovation process, team composition, and leadership roles for the team.

- In particular, a Third Way team requires a broader role for the project leader. The project leader must act as a solution integrator, and have the responsibility and authority needed to lead a diverse team of internal and external personnel through the four decisions.

4

Decision 1

What Is Your Key Product?

DECISION 1 What is your key product?	DECISION 2 What is your business promise?	DECISION 3 How will you innovate?	DECISION 4 How will you deliver your innovations?

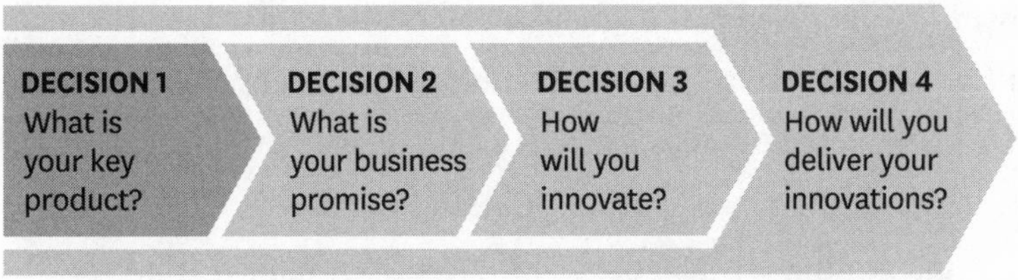

Imagine you've been offered a job running an important subsidiary of a major company. You'll be expected to deliver a profit year after year even though you can't change anything—nothing at all—about the one key product your organization produces. Not only that, but if you were to start turning out a better product, the parent company would reach down and take your best assets. Would you accept that job?

You might if you were a baseball fan. This is exactly the challenge faced by the general manager of the Lehigh Valley IronPigs, a Triple-A farm team of the Philadelphia Phillies. The IronPigs—the name refers to molds used in making steel—have figured out how to succeed under those conditions. Between 2008—when they began playing in a new ballpark—and 2015, they have filled on average over 80 percent of the ballpark's total seating capacity, fixed and temporary, for each of 423 regular-season home games.

How have they done it? Through the Third Way. They can't innovate in their core product, a professional baseball game with its immutable set of rules, so they've had to attract fans by innovating around the game.

By coming up with inventive promotions, novel merchandise, and entertaining stunts, the IronPigs have made a compelling proposition of what has often been a subpar product. Despite a steadily declining record over the last three years, fans return year after year.[1] In fact, the Lehigh Valley IronPigs have the highest average attendance of any club in the minor leagues, despite playing in one of the smallest markets. Examples of promotions and stunts include giving away foam fingers on Prostate Cancer Awareness Night (remember, their fans are mostly men) and free funeral services on Celebration of Life Night. They were the first team in North America to adopt a video game system called Ski the Piste in which the user of a urinal guides a skier down a slope, avoiding fences and aiming for penguins.* FoxNews.com ranked IronPigs fans the best among all minor league baseball teams.[2]

Large organizations typically don't have to live with the stringent constraints the IronPigs face. Most firms have the resources and freedom to innovate wherever and however they want. Given that autonomy, it seems counterintuitive to ask them to begin an innovation program by specifying where they will not innovate.

But a quick look at the companies we've already discussed reveals that this is precisely what they did, even in the face of competitive threats. CarMax's key product was late-model used cars, and Apple's was PCs, at least in the late 1990s.[3] LEGO, in fact, almost went out of business when it tried to innovate away from its key product, plastic-brick-based

* In spite of what it seems to mean in this context, *piste* is French for "ski trail."

construction sets. Only when it returned to the brick—but in a new way, surrounded by complementary innovations—did it go on to renewed success.

In this chapter, we will first describe key products in more detail—what they are and the considerations you should keep in mind as you pick them. Then we will describe, once you've picked a key product, what you must do to prepare it for Third Way success.

Choosing a Key Product

Choosing a key product is a decision to focus on that product, to honor it, and to build around it. It sends a signal to the organization that this product and its customers will be a central focus of effort in the months and years to come. It's also a strategic decision about the type of innovation you won't be investing in: you won't be creating new products that will conquer new markets, and you won't be using "disruptive" technologies to create revolutionary new products that replace existing ones. Instead, choosing a key product is a decision to keep that key product essentially the same and to build a portfolio of complementary products and services around it.

Because a key product or service will be the stable foundation on which you will build an entire system of interrelated innovations, it needs to be chosen with care.

What Qualifies to Be a Key Product?

A key product is any product you've chosen to make more appealing by building a system of complementary innovations around it. Designating a key product is often an easy, straightforward decision, an obvious choice, especially for firms with a strong brand. It's the business you're in, what your customers have come to expect from you. It's what your firm did yesterday, does today, and most likely will do tomorrow.

While we've said it can be any major product or product line, a strong key product usually passes the following six tests:

Is it a product that is or could be central to your company—something that defines or helps define you and what you do? Is it something customers can link with your company and your brand? Will it build or reinforce your company's strategic identity? Will selling it take you toward your strategic goals? Will it build your presence and brand in your company's existing markets or some future market it seeks to enter? In short, is it a good fit with your firm—with what your organization does, how it's seen, and where it wants to go?

Does or could this key product appeal to a distinct and sizable set of customers? The most successful key products appeal, or could appeal, to an important market segment. Without a clear target market, you will have difficulty identifying the needs that the product and its complementary innovations will meet. Such a clear set of needs is the basis for selecting a strong promise and identifying innovations that fulfill that promise (decisions 2 and 3).

Does or could the key product exist more or less autonomously? Or is it somehow tied to another product? For example, it wouldn't make sense for Apple to designate its iTunes Music Store as a key product because it exists only as a complementary innovation that works with and supports Apple's other products such as iPods or iPhones.

Is the key product a stable product? Is it likely to remain essentially unchanged—that is, undergo nothing more than incremental improvements—over at least the next few years? It would make little sense to build a system of complementary innovations around something that will itself undergo significant change in the foreseeable future.

Does the key product draw on and reflect the strengths and values your company is known for? Circuit City chose to launch CarMax largely because selling used cars was a sizable retail business where no one had yet applied the deep expertise in retailing that Circuit City possessed. Apple was virtually alone among computer makers in its ability to make hardware and software work together. By that time in the short history of personal computing, all major companies focused on either hardware or software. Because Apple retained expertise in both, it was able to create a uniquely seamless computing experience for the user. Almost everything it did in pursuing the Third Way reflected and drew on this distinctive capability.

Could the product and its complements generate significant revenues? Handled properly, surrounded by the right complementary innovations, could your key product and its family of complements generate enough profits to justify the effort and resources devoted to them?

All these questions might be summed up very simply: does the product you're considering have the potential for higher sales, and if it produces higher sales, will those sales take your company closer to its strategic goals?

Identify the Key Customers for the Key Product

Once you've identified the key product you wish to innovate around, the next step, a critical one, is to identify the key customers for your key product and why they buy it. The story of Gatorade in chapter 1 perfectly illustrates this indispensable task.

Before Sarah Robb O'Hagan's arrival, Gatorade had been pursuing the often-misguided course we see many companies follow, especially when the growth of an important product has flagged. Product managers search for new customers by spinning out new product variations and new ways and places to market them. Every new variant costs more but sells less. The result may be an increase in sales, but that increase comes with declining profits. Even worse, in their efforts to reach ever more customers in every conceivable market segment, managers lose sight of the segments with the most loyal customers. As they reach for everyone, the product loses its specific and enduring attraction for its most loyal customers.

The decision of Robb O'Hagan and her team to focus on serious athletes took courage. It meant abandoning all those casual customers PepsiCo had added with more product versions and mass distribution. To others at Pepsi, it must have looked as if Robb O'Hagan's response to falling sales was to target a smaller market—not an obvious route to success. To grow sales that way meant selling more to each customer. And that was what she and her team did by adding nutrition products—complementary innovations—around the drink. With those additions, Gatorade, or G, could satisfy a compelling need of serious athletes: it provided the "sports fuel" they needed for peak performance.

The key lesson of Gatorade's turnaround is this: when selecting a key product, choose as well the key customer segment—a "beachhead" set of customers—on which you will concentrate your initial efforts to sell the product. Focusing on that segment and discovering its specific needs will lead you to a more compelling promise, and a more compelling promise will guide you to the most compelling portfolio of complementary innovations that surround the key product. After that initial group of customers has been satisfied, you can expand out from there to additional groups. But the more vague and broad the definition of that initial group of customers, the more vague and unconvincing the promise and all that follows from it will be.

Everything Robb O'Hagan and her people did to turn around Gatorade flowed naturally from their choice of serious athletes as the target customer.

Some Don'ts in Choosing a Key Product

Don't dismiss legacy products. Surprisingly, though the key product in many companies is easy to identify, it no longer commands much attention and respect. It's the old, boring product line that supplies the profits that fund the company's exciting new ventures. Sometimes those old products really are on their way out. But maybe, though, that tired legacy product can be more than just a cash cow. Maybe the Third Way can infuse it with new purpose and life. LEGO thought construction toys based on the plastic brick had seen their day, until it discovered how to revive them for a new generation.

Don't reject a key product candidate just because it's not profitable. This advice may sound odd, but many key products don't make money for the companies that offer them. They're key products because they lead to the sale of other profitable products or services. Auto companies, for example, have been described as manufacturers that make a product that they sell for essentially zero profit in the hopes of getting the financing revenue. Printer makers may sell their product for little or no profit in hopes of selling pricey ink cartridges.

Don't overlook the possibility that a key product may not be yours. This situation is unusual, but there are companies that thrive by selling the complementary innovations that surround someone else's product. To extend the previous example—there are companies that sell discount toner and color cartridges for someone else's printers. Logitech sells accessories—Bluetooth keyboards, laser presentation pointers, mice, and the like—for PCs made by others. Dunhill, the British luxury goods company, began life a century ago by selling "motorities"—accessories such as driving gloves, goggles, horns, and luggage for owners of the then-new automobile.

Don't reject a product because it's struggling in the marketplace. Many companies have turned flawed products into great successes not by improving the products themselves but by surrounding them with the right complementary innovations. CarMax suffers from a purchasing disadvantage with its competitors. LEGO's bricks are far more expensive than competitors' versions. And Novo Nordisk was late to the US HGH market with a me-too product.

Can a Company Have More Than One Key Product?

Avoid the self-limiting mistake of thinking a key product can only be your firm's most important and strategic product or product line. It may be that, and it certainly was for the IronPigs, CarMax, LEGO, Gatorade, and Apple. But it needn't be. For Novo Nordisk, its HGH therapy drug, Norditropin, was secondary to the company's main line of insulin and other diabetes-related products. Still, Norditropin was an important new product with great potential that could only be differentiated by the innovations surrounding it.

If your company is organized around multiple strategic business units or divisions, you should think about key products in the context of those units, rather than the firm as a whole. For example, a car company may, and probably should, consider its family sedans, pickup trucks, and sport coupes to be separate key products (actually, product lines) because they attract different buyers and the complementary innovations that appeal to each set of buyers are likely to be different.

The possibility of multiple key products raises an important question, however. Can you have too many key products, each with a different system of complementary innovations? The answer is yes. Each system will require real time and effort from people across the organization. At some point, the total effort will become too onerous for the firm as a whole. Too many key products can become what Jim Collins calls "the undisciplined pursuit of more," which can lead to a loss of focus on the products and customers that are truly important.[4]

A Key Product Can Change over Time

We've emphasized that a key product is where you will not focus any major innovation efforts. Those efforts will go, instead, to the creation of complements around the key product that make it more attractive. Two exceptions to this general rule deserve comment.

First, you can and should continue to make sustaining, incremental improvements to a key product as long as the changes aren't significant—that is, as long as they don't diminish the product's ability to work with the complementary innovations surrounding it. This point may seem obvious, but some change in a competitive product or some other move by a competitor may tempt you to respond in kind. When AutoNation threatened its very existence, CarMax could have followed by adding new cars to its product line or by acquiring existing dealers to expand rapidly. But both of those options would have led it to abandon its original approach. In the end, CarMax prevailed by continually improving its approach without changing it fundamentally.

Second, a key product and its complements can change because of market response. When Apple released the original iPod in 2001, it was a complementary innovation intended, with iTunes, to drive sales of the Mac computer, which would be the hub of a user's digital life. But that approach changed dramatically in 2003, when Apple launched iTunes for Windows, which effectively made the iPod a key product rather than a complementary innovation. With the later release of the redesigned iPod, the iPhone, and the iPad, those too joined the Mac as key products. Customers could use one or all of them to help manage their digital lives. In 2016 the iPhone, which began almost as an afterthought, an accessory designed to sell more Macs, provided over 60 percent of Apple's revenue.[5]

Preparing a Key Product for Success

Picking a key product is the first crucial step in pursuing the Third Way. But before proceeding to decision 2 (choosing a promise for the key product) and beyond, you must take some further steps in decision 1 to help ensure success. Selecting a key product is akin to doubling down in blackjack, a process where you double your bet, limit your ability to take new cards, and—if done intelligently—increase your chances of winning.[6] Just like doubling down, choosing a key product means you'll be betting more on less, so you should be sure that this decision is made decisively and transparently and that you communicate it widely through your organization.

Identify the Key Product and the Approach Clearly

Because the Third Way will involve so many people across your organization, it's important to begin by consciously and clearly communicating to all involved inside and outside what you've done and what it means. At this stage, you need to communicate three messages consistently.

You need to explain that the product you've chosen will receive special attention. If this product is not the company's core product or an obvious candidate for attention, you also need to explain why this product was chosen—that is, why you think an opportunity exists to increase its sales.

You should describe how most innovation efforts will be focused not on the product itself but around it in the form of complementary innovations that make the product more appealing. Otherwise, people will naturally assume that innovation efforts will focus on the product itself.

Finally, explain that those complementary innovations will come from across the organization and from outside partners, not just from people in product development and marketing. It's important for all groups to know that they can and should play a role.

Make the Key Product Lean and Trim

You may have noticed that every company we've described so far—except for Novo Nordisk, whose HGH drug was a new product—took the same step early in the process. That is, they all rationalized and trimmed the product line they'd just selected as a key product.

Steve Jobs eliminated dozens of Mac models that had been added to the line haphazardly. In focusing on serious athletes, Sarah Robb O'Hagan cut back drastically the number of flavors and variations that Gatorade had added in its quest to become a mass market beverage. LEGO, when it returned to the brick, cut its inventory of shapes and colors in half. And CarMax, after deciding to sell used cars, trimmed its product line even further to focus only on late-model used cars.

What makes all this culling necessary is simple, though possibly painful: if you're going to develop complementary innovations, you will need to coordinate different innovation efforts across diverse groups. That becomes much more difficult if your key product is complex and confusing and its target customers no longer well defined or understood. Also, by reducing the money-losing proliferation of product variants, you will free up resources to invest in complementary innovations.

Make the Key Product as Strong as Possible

As you focus your innovation efforts around a key product, people may assume that all efforts to improve the product itself should cease.

So it's important to be clear: a key product must be as strong as possible. Make sure its production, delivery, sale, and distribution processes are as efficient and effective as possible. When CarMax identified newer used cars as its key product, it worked hard to be better than anyone else at knowing which cars to buy, at what price, in the wholesale market.

When Apple reduced its product line to four Mac models, it began work on a better operating system. It also increased quality and cut costs by outsourcing manufacturing and completely reworking its supply chain. Similarly, when LEGO returned to the brick, it reworked its supply chain and outsourced some manufacturing operations.

The Challenges of Decision 1

Decision 1 can seem so simple and self-evident that you might think making it can be done quickly. This is sometimes the case, but take care to get your key product or service clear, focused, and strong. If you don't, you won't be able to get the subsequent decisions right. Everything else depends on it.

Though it can seem simple, making decision 1 can in practice offer certain challenges. The first, which we described in chapter 3, is that the choice of a key product, the choice of key customers for that product, and decisions around culling a product line can create winners and losers inside your company. Thus, it's important that you make these choices in ways that are both inclusive and open. By inclusive, we mean a process that gives full and thoughtful consideration to all product candidates, so that supporters of "losing" products, though disappointed, will feel they've been heard and the process was fair. By open, we mean a process in which all involved know from the beginning how and by whom these choices will be made.

The second challenge in decision 1 is that making any important product off-limits to change can cut against the grain of today's thinking about innovation, thinking that says everything should be considered a candidate for drastic change. Conventional wisdom says that the best response to a significant threat is to radically change yourself. But we're saying, before you do that, you should first ask, "Do we really need to fundamentally change our important products or services?" Often the answer is no; what's needed is to return to, honor, and build around the product or products that made your company great. Focusing on one or a few key products takes courage and discipline, but many companies have succeeded by doing it well.

When you've completed decision 1, you will have identified the key product—the product around which you will focus your innovation efforts, you will have identified the key customers for that product and you will have made that product line as trim and strong as possible. In chapter 5, we will take you through the next step: identifying the promise you will make to your customers—the promise that the key product and its family of complementary innovations will satisfy.

Three Takeaways for Chapter 4

- The Third Way begins, first, by clearly identifying the key product around which you will create a family of complementary innovations and second, by identifying the "beach head" customer segment—the group at whom you will target the key product.
- A key product should be stable, strategically important for your company, and capable of generating sizeable revenues. It should be a product that you produced yesterday, are producing today, and will continue to produce tomorrow.
- The first step for many companies after selecting a key product will be to trim and strengthen that product. Unnecessary product variants should be culled and innovation efforts refocused on the development of complementary innovations.

5

Decision 2

What Is Your Business Promise?

DECISION 1	DECISION 2	DECISION 3	DECISION 4
What is your key product?	What is your business promise?	How will you innovate?	How will you deliver your innovations?

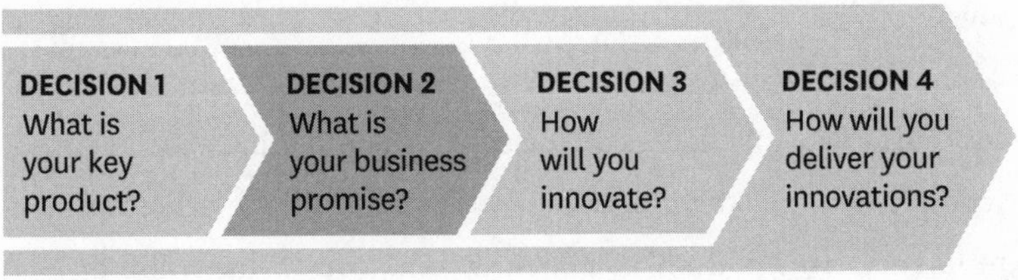

When GoPro launched its first video camera in 2006, it wasn't obvious that the world needed another consumer camcorder.[1] Long-established and well-known companies—including Canon, JVC, Panasonic, and, above all, Sony—dominated that market. It was even less clear that a new camera, in particular one that was technically inferior to the competition, had any chance of success. Yet, less than a decade later, with annual revenues over $1.6 billion, GoPro sold more video cameras—6.6 million—than anyone else.[2] How did it do that? It found a unique promise—to help customers capture their greatest adventures—and used that promise to drive its portfolio of complementary innovations.

Every organization pursuing the Third Way will face a critical decision after it has defined its key product or service. "How will our customers get value from this product?" The answer will decide the innovations you pursue and the success you achieve. Choosing a compelling, specific, and energizing promise will help you select the right innovations to deliver that promise. Choosing the wrong promise, as we'll see, can doom the product line and even the entire company.

The worst option at this stage is make no decision at all. Many companies proceed straight to identifying specific potential innovations, everything they *might* do. Then they select and pursue the best of these. Though some innovations may succeed, the overall results of this scattershot approach are likely to disappoint because, all together, they don't add up to much—just a haphazard collection of this and that, a potpourri of new stuff.

For the Third Way to work as we've described, the individual innovations must work together as a coherent system in which the parts not only complement the key product but also build on each other for maximum impact. This will only happen if some overall intention or goal serves as the glue that binds them to each other and to the key product.

Consequently, decision 2—what is your business promise?—focuses not on which complementary innovations you will implement but on the selection of the innovation promise, the glue, that will link and align all of them. The story of GoPro clearly illustrates the importance of this step.

GoPro Action Cameras

Founded in 2002 by Nick Woodman, a surfer who wanted to capture his experience of riding a wave, GoPro launched its first camera in 2004, a waterproof still camera that you could strap to your surfboard.[3] As of this writing, the company offers a line of affordable, rugged, waterproof cameras ranging in price from $129.99 to $499.99.[4] Most important, GoPro offers accessories that make adventure recording possible: portable power packs, smart remotes, hand grips, memory cards, repair kits, and, above all, mounts for nearly all settings and occasions.* Using the appropriate mount, a surfer can wear the camera or mount it on a surfboard as he or she rides a wave, a family can enjoy again and again its volleyball game on the beach, a biker can capture a ride over rough terrain, and on and on. GoPro removed the bane of traditional photography and video recording—that the person recording the event had to be a spectator filming the action from the outside. Now, with GoPro's cameras and complementary innovations, the camera can come along and record your adventure as you experience it.†

Early adopters, of course, were extreme sports enthusiasts, but word spread and users quickly included amateurs of all kinds, as well as special users such as the US military, police forces, rock bands, and professional sports teams.

But GoPro delivered even more. It recognized that its users wanted to go beyond recording and reexperiencing their adventures; they wanted to record, replay, *and share them with others.*

And so, a key part of GoPro's promise and appeal was that it made sharing adventures easy. For that purpose, it provided two free software packages. The first, GoPro App, lets you control your camera remotely, play back and share the adventures you recorded, and watch "best-of"

* Our favorite: a user-created mount that mounts a GoPro camera on a hula hoop, giving a hoop's-eye view of the hula hooper. See "DIY Hula Hoop Mount for your GoPro," DIYGoPro, December 20, 2013, www.diygopro.com/diy-hula-hoop-mount-gopro.

† GoPro also designed a wrench to help customers tighten the bolts used to attach GoPro cameras to the different mounts. The wrench doubles as a bottle opener, helping customers not only to prepare for an adventure but also to celebrate after.

videos on the GoPro Channel on YouTube. The second free package, GoPro Studio, lets you create GoPro videos, set to music if you wished, on your desktop or laptop and share them on the GoPro Channel. And GoPro integrated its different software packages, making it very easy to move finished videos from the PC to a smartphone, an important step for the millennial market.

To encourage sharing, GoPro offered prizes worth $500 to $5,000—a total of $5 million each year—to creators of action-oriented content that could range from extreme sports to trick shots, talented pets, unusual locations, family adventures, and more. By 2016, users had downloaded the GoPro App twenty-one million times. And, on average, the number of videos uploaded to the GoPro Channel from GoPro Studio had risen to fifty thousand each day.

By mid-2016, GoPro had shipped a total of eighteen million cameras and was represented in more than forty thousand retail outlets in more than one hundred countries. As the company noted, "Our customers include some of the world's most active and passionate people. The volume and quality of their shared GoPro content, coupled with their enthusiasm for our brand, are virally driving awareness and demand for our products."[5]

Sony and other leading makers of video cameras did not watch passively as all this happened. They either created or strengthened their own lines of action camcorders. In many cases, their cameras were technically superior. Sony's action video cameras, for example, were often better in such key areas as picture sharpness, image stabilization, audio quality, and GPS capabilities.

Though Sony and other competitors improved their cameras, they failed to match GoPro's range of complements that made real action recording and sharing possible. Their line of camera mounts, which were critical for action videography, were inadequate and hard to find. They offered little or inferior software for preparing and exporting videos. And their content-sharing communities couldn't come close to GoPro's. Sony's YouTube channel had fewer than one hundred thousand subscribers, while GoPro's had more than three million.[6] Furthermore, even when these competitors offered video editing software and social media outlets, the connection between the product and the service was difficult and error-prone.[7]

As a result, GoPro's 2015 revenue of $1.62 billion represented a five-year average annual growth rate of 91 percent, while revenue at Sony's Digital Imaging Products group declined at a rate of 14 percent.[8] For roughly the same period, GoPro's share of the action camera market was 42 percent, compared with Sony's share of 8 percent.[9]

After establishing itself as a leading maker of action cameras, GoPro seemed to be moving toward a new phase in its development as a company. Sales of its latest camera model were disappointing, in part because of pricing mistakes. GoPro's sales, profits, and stock price declined, and in response, the company has been trying to expand the market for action cameras to include those who want to capture family adventures. Its website and the GoPro Channel have more than ever highlighted videos with family content, it now targets many of its

FIGURE 5-1

GoPro sales and units shipped

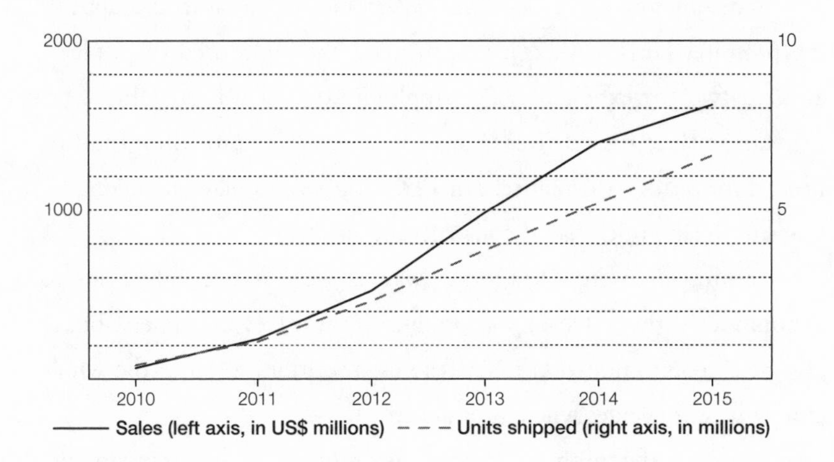

Source: GoPro SEC filings

contests at families, and its marketing efforts increasingly feature family vacations and pets.

In addition, and perhaps more significantly, GoPro is actively moving to position itself as a media company by packaging and licensing in various ways the best of its vast library of video content that users have exported to the GoPro Channel. "The camera is just the tool to get to content," said Adam Dornbusch, GoPro's head of content distribution.[10] If it succeeds in this new effort, it will offer another example, like Apple with the iPod and iPhone, of developing what began as a complementary innovation into a whole new line of business.

Whatever GoPro does now and whatever the outcome, its remarkably successful first decade illustrates the appeal of a compelling promise that addresses an important desire of target customers. Its success reminds us that in the Third Way, it's not the key product alone that is attractive but the product surrounded by complements that fulfill the promise of the product. It is this system of key product and complements that matters, and the promise is what turns it all into an appealing proposition. This promise is the heart of the Third Way and the source of its potential as a competitive weapon.

Every company we've described so far has had a clear and appealing promise. For CarMax, it was to make buying a used car a pleasant, trustworthy experience. For Apple in 2001, it was enabling users to manage their digital lives. For Gatorade, it was providing the fuel needed for peak performance. For LEGO, it was to play out a compelling story using brick-based construction sets.

A promise is useful in another way as well. It not only helps to communicate the value of the product outward to customers, but it also helps communication inside the organization, guiding the whole company as it decides where to innovate.

If you get the promise wrong, the results can be catastrophic. Frederick's of Hollywood and Victoria's Secret both approached the same market with similar products, but had very different promises. These different promises led to the generation of very different complementary products. The result was disastrous for Frederick's, a story we'll come back to later in this chapter.

Our aim in this chapter is to make clear what a promise is and how teams can go about choosing one by studying buyers and their experience with your product. We will outline the steps that help you identify possible promises and the tests that help you choose a good one to pursue. This is a well-studied area that we will summarize and help you explore further by providing references for further study.

A Closer Look at the Promise

A promise is a commitment that you make to your key customer: "If you buy our product, it will . . ." It helps your customer understand how to purchase, use, and get value from your product. It also commits you internally to deliver on that pledge.[11] Strong, appealing promises share some basic characteristics.

BASED ON A COMPELLING CUSTOMER NEED. First and above all, a promise should be based on some deep insight into the basic needs of key customers, especially needs not satisfied by others. The CarMax promise of a hassle-free and trustworthy car-buying experience was compelling for those who needed a used car. Gatorade recognized athletes' need for nutrition and hydration to perform their best. Basing a promise on a compelling need is the best way to ensure that it will be a stable base to build on—customer needs change much less rapidly than technological capabilities or competitor shortcomings, two other commonly used drivers of innovation.

SPECIFY A DIRECTION BUT NOT SPECIFIC INNOVATIONS. In the Third Way, a promise is a lens that helps you focus your search for innovation. It says *where* you will go to find innovations rather than *what* you will do. It acts as a test or standard against which possible innovations can be assessed; those that don't fit are discarded. Its purpose is to focus effort on the area where your key product is most likely to succeed with your key customers.

Like a corporate strategy, which is a statement of where and how a firm will compete, a promise says where and for what purpose you will innovate. A promise does not specify what you will actually do—those choices you will make in decision 3, where they will be guided and focused by the promise you choose here.

CREATE A COMPETITIVE ADVANTAGE. The strongest promises set their products, and the companies that produce them, apart from competitors and competitive products. The customer need they satisfy is one that no competitor satisfies or satisfies as well. Again, every example of the Third Way we've provided matches this characteristic, and if any one of these companies suddenly disappeared, its customers would have trouble finding other ways to satisfy the need it filled.

REFLECT AND ENHANCE YOUR BRAND. Key product promises must be consistent with your corporate brand, but a promise is more specific. We see brands as externally directed statements of what the customer should expect to receive when buying and using a company's products. A promise is more specific than a brand and provides more direction, both internally and externally. A promise focuses on the customer's perspective and gives the company a clear and specific filter to use when choosing complementary innovations. Like a brand, a promise should also be stable, but as the key product and its key customers change, so will the promise.

Procter & Gamble Reexamines Its Promise for Pampers

Procter & Gamble is a well-known consumer brand that the company has worked very hard to connect to quality, safety, and effectiveness. The products that P&G sells—such as Tide, Ivory, and Pampers—benefit from the power of this brand, and the product features and marketing messages for those products have to be consistent with this overall brand. But the overall P&G brand isn't specific enough to be used as a promise—it's too general to guide the generation and selection of complementary products and services. So in 1997, when sales of Pampers were suffering from global competition, Pampers team members developed a new promise that provided the direction they needed to innovate.

In the 1990s, the Pampers promise was clear and simple: Pampers keeps your baby drier.[12] This message had served Pampers well during the early days of disposable diapers, when the comparison was between a disposable diaper and a cloth diaper. But as competitors entered the market, many other disposable diapers kept babies just as dry, at lower cost. The promise still focused on an important customer need, but it no longer created a competitive advantage. The Pampers market share declined.

In 1997, Jim Stengel took over the Pampers brand globally for P&G and began an intensive effort to create a business promise that would guide the development of Pampers products, services, and marketing messages. He and his group did this by immersing themselves in the lives of new mothers to understand their concerns.

What they quickly saw was how narrow their view of their market had become. Mothers cared about whether their babies were dry and comfortable, but cared much more about their children's healthy development and growth. The Pampers team realized that its diapers and complementary products such as wipes did more than just keep babies dry—they kept babies more comfortable and helped keep their skin healthier. All this facilitated the infant's mental and physical development.

Pampers developed a promise for its product: it would partner with moms in their babies' development. This promise guided its website design, promotional giveaways, product positioning, and even the choice of lotions in baby wipes, a complementary product. Advertising emphasized how the product helped a baby sleep through the night and how good that was for a baby's health and development. The team began reaching out to hospitals to provide take-home packs for new mothers. The giveaways contained diapers and wipes and explained how to keep babies healthy and happy, as well as how to choose the diaper that would fit the baby best. The company even partnered with UNICEF to donate vaccinations to babies in developing countries—one for every pack of Pampers sold.

The result was a complete recovery of the business. Market share rebounded, and sales more than tripled over the following years. A strong promise gave the Pampers team the guidance to reposition its products, focus its marketing efforts, and create complementary products and promotions that revitalized the brand.

The Customer Context—Key to a Compelling Promise

The essential foundation for picking a strong promise is a deep understanding of your key customers and how and why they use your key product. The challenge is that learning from customers is difficult. When asked, "What do you want?" they typically do little more than recommend what they've seen in competitive products (see the sidebar "Common Field Research Mistakes" at the end of this chapter).

Why does this happen? Because customers typically cannot see beyond their own experience and cannot envision other ways a product could be used. They rarely know about new technology or other factors that might create new possibilities for the product. And even if they do, it can be difficult to envision how existing technologies and components can be combined in new ways. As a result, customers usually mention incremental, me-too improvements, rather than anything new and daring.

Nonetheless, learning from customers is critical to Third Way success. And so we begin with guidelines for the kind of customer learning that will help you choose a powerful promise.

Learning from customers requires you to go well beyond simply listening. To do this, broaden the scope of what you're trying to understand. Instead of focusing solely or mostly on the customer, seek to understand the full context in which your customer recognizes the need for your product, then finds, buys, uses, and finally disposes of it.[13]

The value of any product or service is not some sort of inherent worth, but is what it adds *in a specific setting*. As the architect Eero Saarinen said: "Always design a thing by considering it in its next larger context—a chair in a room, a room in a house, a house in an environment, an environment in a city plan." How you value a chair, for example, cannot be determined without knowing where it will be used. A plush lounge chair has little value in the context of a kitchen, while a kitchen stool has little value in your office. In the same way, you cannot understand your customer's need for your product without understanding that customer's world and how and where your product fits into it.

Understanding intimately the human context in which a product is used can help reveal not only the most compelling promise for that product but also the complementary innovations that will make it even more appealing. Without exploring the context in which athletes used its product, would Gatorade have found their need for nutrition products before and after exercise?

After studying the companies that do this approach well, we've seen four best practices that we urge you to adopt.

ADOPT A "DATING, NOT FIGHTING" MINDSET. Developing this deep understanding requires a mindset that can cut against the grain of thinking in many organizations. Too often, those responsible for innovation focus on the competition and how to beat them. They seek to out-innovate the enemy. The problem with this way of thinking is that it leads you to focus on what the competition is doing rather than what the customer needs.

The better approach was summed up by Bob Wells, senior vice president of communications for Sherwin-Williams: "We've always looked at business more like dating than like war . . . In war, you're focused on beating the competition. In dating, you're focused on strengthening a relationship. That difference in perspective has a million knock-on effects for how decisions get made."[14]

TO UNDERSTAND TIGERS, GO TO THE JUNGLE, NOT THE ZOO. The need to understand the context in which a customer uses your product suggests another important practice: going to that world and observing customers as they actually use your product. Thoughtful observation can lead to insights into customer needs that customers themselves don't yet recognize or cannot yet articulate.

LEARN FROM YOUR MOST EXTREME CUSTOMERS. If you are able to identify heavy users of your product, compare their responses with those of regular users. Are the heavy users attracted by some product feature that others usually overlook? Are they using the product in unusual ways that might appeal to others? Look for ways of growing regular customers into heavy users.

The reverse of this piece of advice is also true: talk to and observe potential customers who don't use your product. Why don't they? What gets in the way? Sometimes, talking to these "virgins" can show you barriers to adoption that you didn't know existed.

WHEN IN THE FIELD, PRACTICE "VUJA DE." Déjà vu is the feeling that you've been somewhere before, even when you haven't. "Vuja de" is the opposite—going to a place you've been many times, but seeing it with fresh eyes.[15] By doing so, you'll see the annoyances and frustrations that your customers have learned to live with and adjusted to. Each of these irritations represents an opportunity for improvement.

There are three steps in this second decision (see figure 5-2). As with the overall Third Way process, we depict these steps as a sequential flow, but in practice you may find yourself going back and forth between them. The first step is to map the customer context.

FIGURE 5-2

Finding your business promise

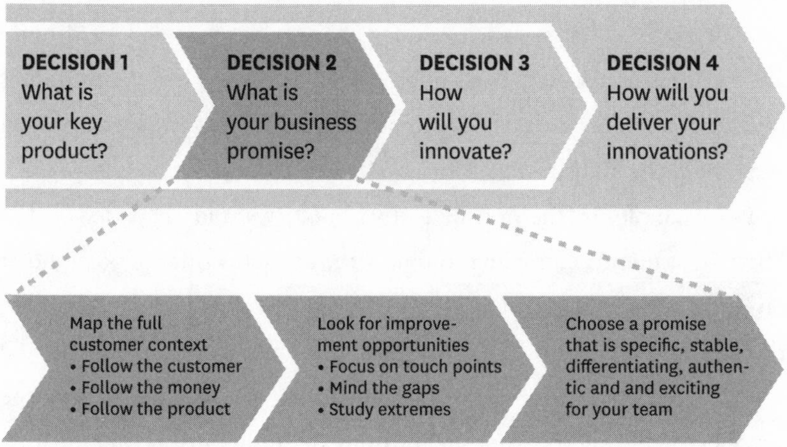

DECISION 1
What is your key product?

DECISION 2
What is your business promise?

DECISION 3
How will you innovate?

DECISION 4
How will you deliver your innovations?

Map the full customer context
• Follow the customer
• Follow the money
• Follow the product

Look for improvement opportunities
• Focus on touch points
• Mind the gaps
• Study extremes

Choose a promise that is specific, stable, differentiating, authentic and and exciting for your team

Step 1: Map the Full Customer Context

The first step in finding a great promise is to map the three chains of activities that will illuminate the customer context around your key product: (1) the way your key customers *use* the product, (2) the way they *buy* the product and other related items, and (3) the way *you deliver* your product or service—that is, the activities you undertake to conceive, design, make, distribute, sell, maintain, and even perhaps help dispose of the product. We've labeled these three chains, respectively, *follow the customer*, *follow the money*, and *follow the product*. When studied together, all three can reveal entirely different insights about your customers' needs and how you are meeting or might meet them.

We think of this three-phase analysis as unfolding in two dimensions, as shown in figure 5-3. Two of the activity chains (shown crossing diagonally in figure 5-3) are related to your customers' efforts to find, buy, use, and get value from your products. Once these are understood, you'll need to analyze and improve your own activity chain—the set of activities to produce and deliver your product to the customer.[16]

For some types of products and services, following customers as they use the product will be much more productive; for others, following the flow of funds—that is, where money changes hands—will yield more and better insights. And analyzing how, where, and when your product touches (or could touch) the customer in the first two chains can reveal even more possibilities.

For example, in the preface to this book, we told the story of The Sherwin-Williams Company, and how the company offers a portfolio of products and services that make its paint more valuable to its customer, the painting contractor. Following the customer would document the activities such as planning the job, scheduling the work, acquiring the supplies, preparing the work site, priming, painting, inspecting the final

FIGURE 5-3

The three activity chains

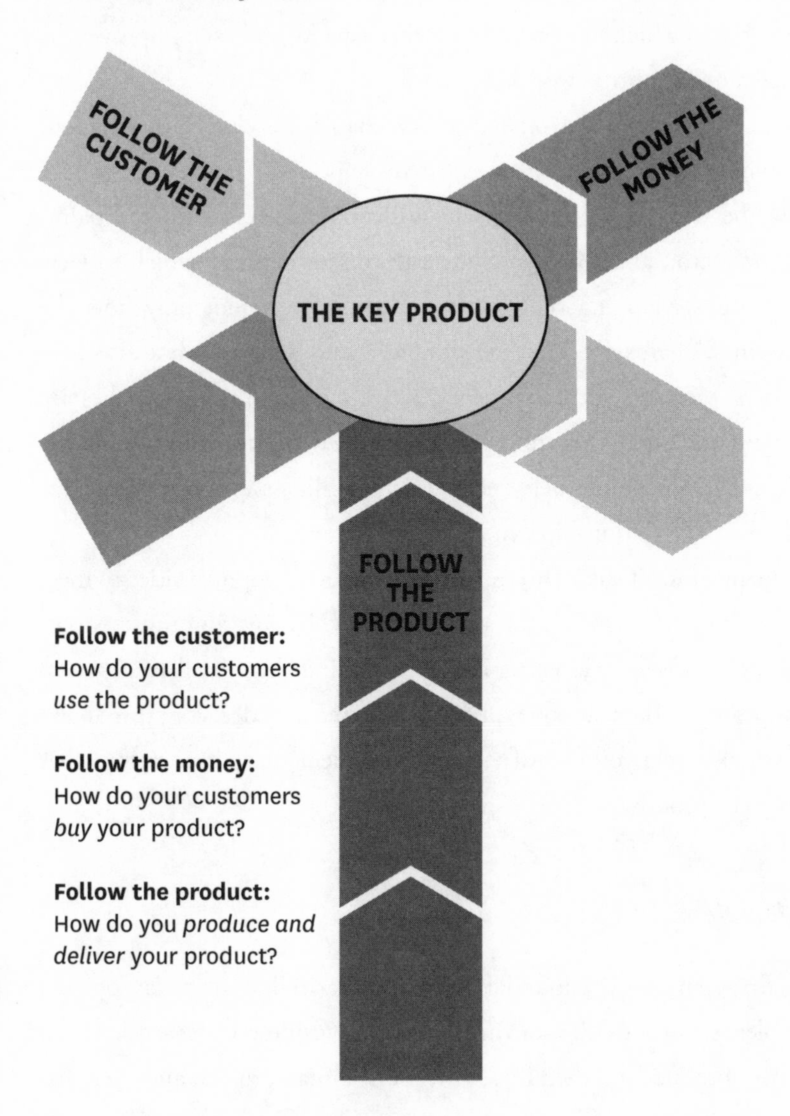

result, and cleaning up after. Following the money focuses on the transactions where funds are promised or exchanged. Analyzing this chain of activities would show the process of bidding the work, agreeing on a price, hiring labor, acquiring supplies, completing the job, getting paid by the homeowner, and paying the workers. These two activity chains intersect at different times and in different ways, but each provides a lens through which to view the customer's world, and each improves the understanding of that world.

Once those two customer activity chains are clearly understood, following the product, the vertical dimension in figure 5-3, would show the ways in which Sherwin-Williams connects with the painting contractor, and would provide a third lens through which to view the process. Mapping these activities would show not only how the Sherwin-Williams products are produced and delivered, but also how partners' products such as brushes, tarps, tape, and other supplies are acquired and sold. This third lens is especially helpful for showing the different touch points between Sherwin-Williams and the contractor and how that relationship can be improved.

The purpose of these three analyses is to reveal opportunities to meet customer needs in some new and better way. Uncovering those opportunities, and then evaluating and synthesizing them, will lead you to your promise. Then, as you will discover in the next decision, this analysis will also help you identify specific complementary innovations that satisfy the promise.

Follow the Customer

The first step and the foundation for all that follows is to lay out the complete set of activities starting when your customers first sense a need for your product and then find, buy, use, maintain, repair, and then dispose of it. This *customer activity chain* is the core of the customer context.

The key to following the customer is to look behind your customers' activities and understand what they're *trying to accomplish* with your product at every step along the way as they find and use it. Customers often cannot express, explain, or even know what they want or need.

How can they, if they don't know what's possible? But they can tell you something even more useful: the job they're trying to do, the outcome they're trying to accomplish, and the result they want to achieve, when they use your product. Once their overall goal or job has been identified, every step in the chain of activities leading up to completing that job can be evaluated in light of the ultimate result being sought.

This jobs-to-be-done approach, as it has come to be called, arose from Theodore Levitt's still-relevant *Harvard Business Review* article "Marketing Myopia," which first appeared over a half-century ago.[17] In it, Levitt argued that people buy a product not for itself but as a means to accomplish some task. The now-classic example of this concept is that people don't buy a drill because they want a drill; they buy a drill because they want a hole.[18]

The appeal and power of this approach is that many products force the user to compromise because they don't do the complete job the user wants done. That's frustrating because it forces users to cobble together their own solutions around the product. The Third Way is an ideal method for dealing with this situation. By knowing what outcome your key customers are trying to achieve, you can begin to discern needs they cannot yet express directly. By adding complementary innovations around the product, you can provide a solution that does the whole job.

GoPro's early success can be traced to its deep understanding of what customers were *trying to do* with its products. From the process of mounting the camera, to planning which parts of the experience to record, to editing the video, GoPro has developed an integrated set of products, accessories, and tools that support the entire process (figure 5-4).

To document the customer activity chain, break down the user's activities into separate steps, describe each, and then ask users what job they're trying to accomplish with each. Finally, question the user to gather two additional pieces of information about each of the steps in the chain. First, rank the importance of each step to getting that job done.

FIGURE 5-4

GoPro's success in following the customer

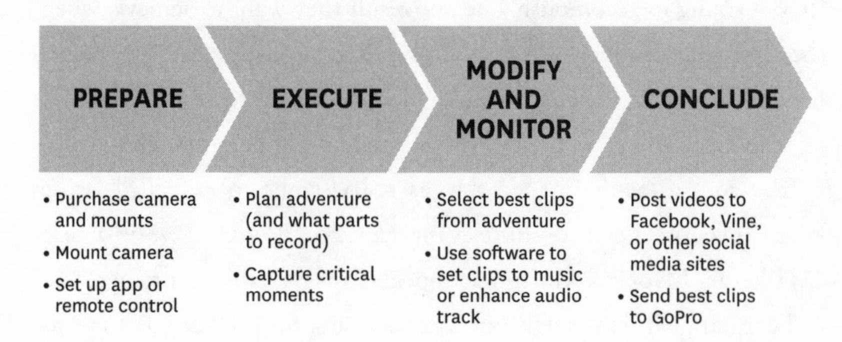

Each step can range from crucially important to entirely optional. Second, rank how satisfied the user is with the way that step is getting done now. This information will come from observing customers, asking what purpose or job they are seeking to do, and then having them rate each step for importance and satisfaction.[19]

Follow the Money

Once you understand how your customers first develop a need for your key product and then find and use it, the next step is to identify where and why money changes hands. Following the money will map a different but related set of activities—the consumption chain—and will help you understand where value is recognized by any of the parties involved. Obviously, the key point in this chain is where the customer pays for your product. But there may be other places in the chain where the customer pays someone other than you for something related. You encounter this whenever you order something online; once you've placed the product you want into your online cart, the vendor—say, Amazon—shows you several related products and says, "Customers who bought this product also bought these other products." If you buy, for example, a vacuum cleaner, you'll be urged to buy the disposable filters and bags that fit that vacuum.

As with following the customer, you'll want to get out into the field to understand what really happens, but here you're focusing on the flow of funds. Who is spending what with whom? What are all the different transactions that take place over the customers' entire experience with your (or your competitor's) product? What is exchanged for how much at each step? And, as with following the customer, you'll want to rate the importance of each step, and the customer's satisfaction with that step.

In some cases, this approach will lead to great insights that follow-the-customer research might miss. We have found with business-to-business situations and with more complex consumer products, that following the money can yield profound insights.

For example, when CarMax did its initial research, the team would not have learned much from following the customer. Tracking the customer's use and maintenance of the vehicle might have led to ideas for new features for the vehicle, something that would be important for automakers to understand but not for the CarMax management team.[20] Instead, it found that the follow-the-money approach led to the real opportunity. The flow of funds and exchange of goods and services was an intensely frustrating process for buyers of used cars, and CarMax successfully created a business that removed those frustrations (figure 5-5).

While some of the insights gained from following the flow of funds will overlap with those gained from following the customer's use of the product, the two approaches often result in different insights. Each is a unique lens that lets you view the same customer's behavior in different ways.

For a lengthy list of possible steps and some recommendations for researchers, see the work of Ian MacMillan and Rita McGrath,

FIGURE 5-5

CarMax's success in following the money

AWARENESS AND SELECTION	PURCHASE	FINANCING AND INSURANCE	USAGE AND REPAIR
Wide selection of newly used cars; inventory from any other store easily available	No-hassle, fixed-price sales; fixed-price trade-in offer; five-day full-money-back guarantee	Multiple offers and separate transactions for financing and insurance	Sell newer models so that most are still under manufacturer's warranty; extended warranty available

two academics who have spent years understanding how to map the consumption chain.[21]

Follow the Product

In this final step of understanding the customer context, you will map the value chain, the activities performed by your company around your key product—from designing it to producing, marketing, distributing, selling, delivering, supporting, and even helping to dispose of it.[22] As you try to identify possible promises in the steps that follow, this knowledge will be invaluable. It will allow you to identify where your internal processes touch the customer's, and where they don't touch but should. In short, it will help you identify where you currently add value for the customer and where there are opportunities to add even more.

We won't spend time in this chapter reviewing the extensive literature on value chain analysis—it's the most mature and best-known type of analysis. Readers interested in learning more are urged to consult Michael Porter's books or the more recent work of Larry Keeley and his colleagues.[23]

Step 2: Look for Opportunities

Once you've completed your analysis of the three activity chains around your key product—the customer activity chain (follow the customer), the consumption chain (follow the money), and the value chain (follow the product)—the next step is to identify opportunities for improvement, each of which could suggest a possible promise.

Every Step Is a Potential Opportunity

A key insight we've stressed again and again is the value of thinking beyond the product itself. As you study the customer's world, think of every customer activity as an opportunity to innovate, a chance to make your product more useful and appealing and to set it apart from the competition.

Identify Actual and Potential Touch Points

Touch points are places where you, the seller, and the customer interact in some way. Note those places. Identify as well the places where your competitor or competitors interact with your customer.

Laying out all three chains, with interaction points noted, can generate a wealth of insight, much of it unique to your specific setting. We can only suggest some obvious questions to ask as you examine the three chains and how they interact.

- Look at the points on the customer's activity chains where you do not currently connect with the customer. Are any of those points opportunities to connect in some new way that improves the customer's experience and strengthens your relationship?

- Look especially at places where your competition connects with the customer but you do not. Should you create a connection? Can you improve on the competitor's performance at that point?
- Review the points where you already connect with the customer. Are there ways you can improve the buyer's experience at those points?
- As you follow customers through the steps in their value chain, look for other parties— beyond you, the customer, and your competition—who play a role in that value chain. Is that role something you could and should assume?
- As you follow the money, note places where something is bought and sold but where you're not involved. Are there ways you can take part in (or take over) that transaction?

Mind the Gaps

Look now at the two rankings—for importance and satisfaction—assigned to each activity by the customers you surveyed. Look in particular at those activities ranked high in importance and low in satisfaction.

These are likely to be sore spots, sources of frustration and even anger for buyers. Unless you're already involved in that activity, a high-importance/low-satisfaction rating signals an opportunity—if you can find ways to address it. If you are involved already, that combination of ratings is a warning that you need to do better.

If, in surveying customers, you asked the reason for any low-satisfaction rating, you should have some idea of how to improve. If you lack that information, you will need to revisit customers to understand the reason for their unhappiness.

Pay Attention to Major and Extreme Customers of the Key Product

If you were able to identify heavy users of your product, compare their responses to those of regular users. You may be able to find ways of growing regular customers into heavy users. Are these users attracted by some feature of the product that others usually overlook? Are they using it in unusual ways that might appeal to others? As we said earlier, be aware of the possibility that heavy users may be different from regular users in ways and for reasons that only apply to them.

Document Your Work for Use in Decision 3

Much of what you do and learn at this stage will feed directly into decision 3, where you choose a variety of complementary innovations that will make your key product more attractive. In fact, your search for a promise will generate many ideas for those complementary innovations. Record them as they come to mind, and look at them in decision 3. Those that fit the promise you ultimately choose will become candidates for complementary innovations.

There's no magic formula for success in this step—no algorithm that will deliver a great promise. What's needed is careful fieldwork, detailed analysis, and thoughtful reflection. As you uncover opportunities, use them to generate possible promises. And spend time inside your company as well—your own employees will often have ideas for what needs to be improved and how your company can better deliver value to customers.

Step 3: Choose a Promise to Pursue

Once you've generated a list of possible promises, your task is to select one that will guide and unite all further efforts around the key product. There's no quick and easy way to make this selection, but there are certain tests you can apply that will help you determine the strongest possible promise.

> *Is the promise specific enough to help you decide between different innovation ideas?* This is the most important function of a promise. It is a statement of the outcome the key product and its complementary innovations will deliver. It focuses all those pieces on producing that one outcome.
>
> *Can you tie the promise back to the insights you gained in the field?* A strong promise addresses a customer need or desire in a simple, compelling way. For it to do this requires that you know intimately your key customers and the context in which they work or live.
>
> *Does it differentiate you?* The strongest promise sets you apart from your competitors. The customer need it addresses is ideally one that so far has gone unsatisfied or, at least, under-satisfied. Clearly, it was their business promises that made CarMax, Gatorade, and GoPro different, unique, and better. Seek a promise that will do the same for your organization. Best of all is a promise that sets you apart in ways that will be difficult for competitors to match. You need to know what competitors are doing and where you may need to match them, but a promise that simply delivers what competitors already deliver is unlikely to take you far.

Will it be stable over time? You want a promise that is unlikely to change soon, one that will remain compelling for some time. Otherwise, the effort you devote to creating complementary innovations will probably never pay off. This is yet another reason to derive your promise from some basic customer need because a basic need is unlikely to change soon.

Does it build on your company's unique strengths? When Gatorade decided to expand beyond hydration and into energy snacks, it could draw on the expertise of Quaker Oats, another PepsiCo company. When Circuit City created CarMax, its plan was to take the deep retailing skills it had already developed and apply them to a new business.

Note also that all these companies' promises reflected their core values. LEGO's promise continued its lifelong emphasis on creative play. Gatorade's new energy products were an obvious way for it to support the serious athlete, its original key customer.

Will the promise excite your team? The final test for a promise is whether it is energizing for your company. Does it excite the passions of your employees? Does it represent a noble and important purpose for your company? Will your company's leaders be able to rally the company behind it? Your promise should represent an aggressive challenge that will transform the company when it's achieved.

Is your key product central to delivering on your promise? Once you've chosen a specific promise or have reduced the candidates to a short list, you may want to return to decision 1 and review your choice of a key product and target customers.

> Is the promise you have chosen or are considering a natural fit with your key product? Is the connection one that key customers will grasp intuitively? Or does it require explanation? In light of the promise, review your decisions regarding which product variants to drop. Do those choices still make sense? Are there others that can be cut, or should some be reinstated?

Once you've settled on your promise, you'll be ready move to the next decision where you select the specific complementary innovations that, in combination with the key product, will fulfill that promise for your key customers. Much of the work you did here, in particular, the customer insights you gained by fleshing out the customer context, will be useful in your search for complementary innovations. This search and selection process is the subject of the next chapter.

The Power of a Promise: Why Victoria's Secret Beat Frederick's of Hollywood

An object lesson in the power of a good promise comes from the women's lingerie industry.[24] Frederick's of Hollywood and Victoria's Secret each offered similar key products: intimate apparel for women. Both created a network of dedicated retail stores and used popular models to represent their brand promise (Frederick's of Hollywood: Pamela Anderson and Brooke Burke; Victoria's Secret: Heidi Klum, Tyra Banks, and Gisele Bundchen). Further, both have successfully innovated in their core product lines: Frederick's of Hollywood invented the push-up bra, while Victoria's Secret created the Miracle Bra.

That, however, was where the similarities ended. While Frederick's, founded in 1947, was the market leader for almost forty years, Victoria's Secret developed a very different promise and drove Frederick's physical stores out of the market. In 2013, Victoria's Secret announced a net income of $753 million, with $10 billion in revenue. That same year, Frederick's of Hollywood reported a net loss of $23.5 million against $86.5 million in revenue. Frederick's was taken private in May 2014 and closed its last physical store in 2015.

Given that Frederick's of Hollywood was the market leader and the two companies initially pursued the same market strategy, this reversal of fortunes is remarkable. Like Frederick's, the Victoria's Secret image in its early years was "more burlesque than Main Street."[25] However, the company was bought out by L Brands in 1983 and quickly changed its focus. While the previous owner had believed that men bought lingerie for the women in their lives, extensive field research by the company revealed that women often found the lingerie that men bought them unappealing and uncomfortable.[26] Victoria's Secret refocused its products, introducing "new colors, patterns and styles that promised sexiness packaged in a tasteful, glamorous way and with the snob appeal of European luxury."[27]

This redefinition continued and expanded in 2000, when Sharen Turney, CEO of Victoria's Secret Direct, changed the racy Victoria's Secret catalog to something closer to a Vogue lifestyle layout, while Grace Nichols, CEO of Victoria's Secret Stores, similarly transformed the look and feel of the stores away from a Victorian bordello to a luxury shopping experience that emphasized romance and passion.[28] Frederick's stores, on the other hand, remained evocative of a boudoir costume shop, with their signature red color scheme and scantily clad mannequins in come-hither poses.[29]

Frederick's of Hollywood amplified this sleazier image when it began offering lascivious undergarments and sex toys as well as push-up bras, panties, and corsets. Victoria's Secret went the other direction, introducing products ranging from swimwear to CDs featuring romantic classical music. In 1989, the company announced its expansion into toiletries and fragrances, released its own line of fragrances in 1991, and entered the $3.5 billion cosmetic industry in 1998.

Thus, although both companies were built around the same core, lingerie, they based their innovation on very different business promises. Frederick's remained lascivious; Victoria's Secret projected refined sensuality and romance.

And Victoria's Secret has not been content to rest on its scantily clad laurels.[30] The company continues to experiment with different types of complements, from pajamas to swimwear and sportswear. In 2005, the company opened its first airport store, in London's Heathrow Airport, and in 2010 began aggressively expanding its network of airport stores. These airport stores, and similar outlets located in tourist destinations, are notable because of what they *don't* sell: the company's lingerie products! The stores are much smaller than the mall stores, and feature Victoria's Secret fragrances, cosmetics, and other complementary products.

Frederick's of Hollywood, imprisoned by a weak business promise, closed its last store in April 2015 and is now solely web-based.[31] Driven by a more popular promise that led to an entirely different set of complementary innovations, Victoria's Secret continues to thrive and grow with an impressive 35 percent market share. A better promise has made Victoria's Secret the only name that matters in lingerie today.

Three Takeaways for Chapter 5

- Your promise communicates the compelling customer need that your key product and its complements will fulfill. A good promise will be specific, authentic, stable, differentiating, and exciting for your team.

- Three overlapping but distinctly different types of analysis will give you a rich understanding of the customer context: follow the customer, follow the money, and follow the product. These three types of analysis will illuminate the customer activity chain, consumption chain, and value chain, respectively.
- Analysis of the three chains will help you uncover unmet customer needs and discover new opportunities, allowing you to generate and test different alternatives for the promise.

COMMON FIELD RESEARCH MISTAKES

Look out for various common pitfalls and traps that snare those looking to understand the customer's real needs and desires.

Don't focus on the competition. Defining your goal as "beating the competition" will lead you to produce a me-too offering when the goal is to satisfy the customer in ways the competition cannot match.

Don't do field research without an interview guide. It's common for most people to ask leading questions that subtly signal an expected or a desired answer. Only with practice and a solid interview guide with open questions can you get to an unmet need while studying a customer's behavior.

Don't be constrained by how customers currently use the key product. Instead, look beyond current use to what customers *could* be doing with it. Enlist noncustomers, and watch them try to use the product, with very little direction. Observe, and ask open-ended questions.

Don't stop with "knowing that." You need to *know that* customers prefer, say, your product in black and not some other color, but it's even better to know *why* they want black. Knowing why customers prefer something gives you specific direction for future innovation efforts.

Don't anchor on current products. Asking customers the strengths and weaknesses of current products tends to elicit information about competitive offerings—features that competitors have that your customers wish yours had. This is useful information, but it will rarely get you to a deeper insight.

Make sure you're not just confirming your own beliefs. If you're not aware of your personal preferences, biases, and inclinations, you're likely to use research to confirm those preexisting beliefs rather than find something truly new and useful.

6

Decision 3

How Will You Innovate?

DECISION 1	DECISION 2	DECISION 3	DECISION 4
What is your key product?	What is your business promise?	How will you innovate?	How will you deliver your innovations?

Inside a nondescript office building in Bellevue, Washington, sits one of the most profitable companies in the tech industry. Valve Corporation's annual revenue per employee—more than $2 million—far exceeds the best performance of Google, Facebook, Apple, or any other publicly traded tech giant. Yet it's a company known only to passionate gamers.

Valve has been able to capture an estimated 50 to 70 percent of its market—the distribution of PC-based games—because it has mastered decision 3. It has continually developed innovative products and services that deliver on its promise of providing the best social entertainment platform.

In this phase of the Third Way, the goal is to generate as many ideas for complementary innovations as possible, choose the best of them, and test whether they'll really work. Decision 3 may seem like familiar territory because idea generation, selection, and testing have been widely studied and extensively covered in the innovation literature. It will also feel familiar because generating and selecting innovation ideas are what many companies consider the essential steps of the *entire* innovation process, steps already well known to them.

But in the Third Way, decision 3 requires some unique twists to familiar routines—twists that can be tricky to manage. In this chapter, we'll discuss these challenges and show how to navigate them. Our goal is not to repeat concepts and approaches that have been well covered elsewhere but to review that material quickly and highlight the changes needed to execute decision 3 well.

The Story of Valve

Valve was founded in 1996 by two former Microsoft employees, Gabe Newell and Mike Harrington. Its first video game, Half-Life, released in 1998, was an immediate hit. Since then, Valve has continued to produce more games, but it has also maintained a relentless focus on improving the entire gaming experience. It has continually studied, measured, and improved the different activity chains its customers and partners go through, and by doing so has become the most dominant force in the PC gaming industry.

One result of this constant focus on gamers and their experience is that Valve has steadily changed its entire focus. Instead of just producing great PC games, the company is moving toward providing a gaming platform that supports the development, marketing, and distribution of games. This wasn't a grand strategic move that happened all at once, but a gradual transition built on a close connection with, and deep understanding of, its customers. Its promise now is to provide a rich, diverse, and trustworthy social entertainment platform.[1] As we saw with Apple, LEGO, Victoria's Secret, and others, this isn't an unusual result for companies that have adopted the Third Way. The focus on delivering the promise often leads companies into new markets.

For example, early in its life, Valve realized that the most frustrating part of the gaming experience had nothing to do with the games themselves. There were lots of good games available. Rather, the *distribution* of games was inefficient, expensive, and frustrating. Mall game stores carried only a small selection of games and marked up prices tremendously. They provided no objective rating of the quality of the games and were often staffed by low-paid clerks who knew little about the business and cared even less about the customer. So Valve started its own distribution channel called Steam, an online platform where gamers could buy and update games. At first, Steam handled only Valve games, but it soon added third-party games, for which it charged a commission.

Valve did something similar with in-game collectibles (IGCs), which are game objects and characters with special characteristics or powers that expert players could acquire with superior play. When Valve saw expert players selling their IGCs to less experienced players, sometimes for thousands of dollars, it created a market for IGCs and charged a commission on each sale.*

And when Valve saw that gamers were upgrading their standard computers with add-on hardware such as faster graphics cards and better controllers, it invented its own operating system, partnered with hardware vendors, and began selling PCs that came optimized for gaming.

*One example: a pink Enduring War Dog with an Ethereal Flame particle effect was sold for $38,000. Valve gets a commission of 15 percent of that, or $5,700. We have no idea what that is, or why it's so valuable. We asked some teenage gamers to explain it to us, and we still don't understand.

It also developed and began selling its own controller that could be customized for different types of game play.

These examples show just some of the innovation that Valve has done to improve both the purchase process and the use process. But it has also invested great effort in supporting the development and launch of new games. Following the product from beginning to end has helped Valve greatly increase the range and quality of the games available and its profits as a result.

One of the first steps in helping other companies develop games was the release of Source, a powerful game development environment. Source gives independent game developers the same set of tools that Valve uses internally to develop its games. Valve began giving away Source for free in 2004. To help developers choose which games to invest in developing, Valve created Greenlight, a site where potential customers could view game trailers, vote on which looked best, and give feedback to developers. Then, Valve created Early Access, a site that lets developers release early versions of their games to small groups of users who test and provide feedback. And Valve recently created a tool called Steam Workshop that allows developers to open up their games to customer-generated "mods" (modifications) and incorporate ideas from those mods into the games.* To help publicize games, Valve created Moviemaker, a tool that lets game developers or players make movies illustrating the entertaining parts of their favorite games. All these tools—Source, Greenlight, Early Access, Workshop, and Moviemaker—are free to members of the Steam community.

With steps like these, Valve has been able to expand around its key product—the game development and distribution platform

*Our favorite mod occurred in the medieval adventure game Skyrim. One particularly creative user created a mod that replaced all the dragons with Thomas the Tank Engine. See the results at https://steamcommunity.com/sharedfiles/filedetails/?id=201861191.

Steam—while remaining focused on its key customers: serious gamers. Valve is constantly following the customer, the money, and the product to ensure that it's innovating around every aspect of the gaming experience.

The following sections describe the process for choosing the complementary innovations you will actually create. The description is necessarily detailed in places, but overall it's simple: The process comprises only three basic steps.

1. Generate a portfolio of ideas for complementary innovations.
2. Narrow that portfolio by applying to each idea some specific filters that we will describe below.
3. Finally, use experiments to choose the complementary innovations you will actually deliver.

Generate a Portfolio of Ideas for Complementary Innovations

Valve has profited immensely by practicing the Third Way. Its employees understand at a very deep level the customer context and have innovated in multiple ways to improve it. As you move from decision 2 to decision 3 in your company, the good news is that you've already begun much of the necessary work. In decision 2, you analyzed the different chains of activities that your customer and you go through to buy and use your products. The purpose there was to create a strong promise for your key product.

Start decision 3 by returning to that analysis, and use it to identify possible complementary innovations that, when deployed alongside the key product, will help you deliver on that promise. When you analyzed the customer context for your key product in decision 2, you mapped out three chains of activities:

1. You *followed the customer* by mapping the customer activity chain—the entire sequence of activities by which your customers prepare for, use, modify, and complete the task that they're "hiring" your key product to do. This step gave you a clear understanding of what the buyer was trying to accomplish with your product and where that chain of activities was difficult or frustrating.

2. You *followed the money* by identifying all the places in the consumption chain where money changes hands. Doing this can reveal different insights because it will reveal where in the consumption chain value is recognized and paid for by the customer or someone else. As you will see, this step is what led Valve to many of the opportunities it seized.

3. You *followed the product* by mapping the activities in your value chain, which are the activities performed by you, the seller, to design, make, market, sell, deliver, and support your key product.

Return now to those processes and, using the analysis we suggested in chapter 5, challenge your team to generate not a promise as you did in that chapter, but as many ideas as possible for specific complementary innovations that satisfy your promise. In addition, team members should apply the best practices that they undoubtedly know already, such as setting up innovation tournaments, holding brainstorming sessions, sending out Kickboxes (see the sidebar "Put Out a Kickbox Based on the Promise"), and involving outside partners. Your goal is to create a rich list of ideas for fulfilling the promise you developed for your key product.

PUT OUT A KICKBOX BASED ON THE PROMISE

Kickbox, an approach to corporate innovation, was developed and used internally by the software company Adobe. It's now available online to anyone who wants to try it (see kickbox.adobe.com). Kickbox is based on a belief central to the Third Way: that innovation can come from anywhere in the organization. Its core is a red box that contains all the material an employee needs to flesh out an innovative idea, validate it with field research, and report back the results.

At Adobe, any employee can request and will receive a Kickbox. Adobe only requires the recipient to attend a two-hour training course that covers how to use the materials and reviews the kinds of projects—the technology areas and market segments—that would most benefit Adobe. Each box also contains a $1,000 prepaid credit card that the innovator can use to explore and validate his or her idea. (Adobe considers the money an essential part of the approach.)

There is no application and approval process. Every recipient is free to explore any idea that is relevant to the strategy of the company. When the recipient has completed the six-step Kickbox process, he or she reports back to Adobe management the idea explored, the market data generated, and a recommendation for next steps, if any. No one is punished in any way if the results are negative or even useless.

Adobe sees at least two benefits from Kickbox. First, of course, some of the ideas are worth pursuing (the company doesn't release information about what worked or not). Second, the company views the whole process as an effective training tool. Even someone whose idea led nowhere has learned how to pursue an innovative idea, is much more familiar with Adobe's strategic priorities, and has been exposed firsthand to Adobe's marketplace and the people in it.

Although Kickbox can work for any type of innovation, we recommend this approach to any company beginning a Third Way program for two reasons. First, it can help increase the range and quantity of innovation ideas available to choose from. Second, and more important, it can help to teach innovation tools and techniques to a broad group of people in all parts of the company, a necessary condition for the Third Way to succeed.

Narrow the List of Possible Complementary Innovations

Once you've created a comprehensive list of possible innovations, winnow the list to those you believe are worth implementing. For that purpose, you need a set of filters or tests for evaluating each one. We suggest using three that we already mentioned in chapter 3—constrain, connect, and control—to which we now add a fourth: *complete.* Use these tests in two ways: first, to remove ideas that seemed great when someone suggested them but that don't actually qualify, and, second, to assess the attractiveness of those that seem to qualify but present differing degrees of risk.

Notice that we did not include an often-used test of innovation attractiveness: will the innovation generate profits for the company? For reasons that will become clear in the next chapter, we don't consider profitability a necessary hurdle for every innovation. What matters is the overall profitability of the system—the key product and the complementary innovations around it. Sometimes, an innovation will lose money but it plays an essential role in making the whole portfolio more profitable. Many marketing innovations, for example, are costly expenses but are happily incurred because they will lead to greater success overall.

Constrain: How Crucial Is the Innovation to Delivery of Your Promise?

The selection of a complementary innovation in the Third Way hinges, first and foremost, on whether it helps the key product and other innovations satisfy the promise. The promise is a constraint on which innovations are acceptable or not. This initial test is the most important: either the innovation helps fulfill the promise, or it's out.

As we said in chapter 3, you will sooner or later come across innovations that, after careful consideration, really don't help fulfill the promise. But they possess other attractions that tempt you to implement them nonetheless. So seductive are they that some will argue to make an exception for them. However, no matter how appealing, these attractive but deficient ideas will require time and other resources that would otherwise be devoted to the Third Way. An innovation that's not needed to satisfy the promise should be dropped, no matter how tempting.

Connect: How Will the Innovation Link to the Key Product and Other Complementary Innovations?

Evaluating whether a complementary innovation connects to the key product and other innovations in the portfolio involves asking a number of questions. Will it need to, and be able to, connect physically to the key product? Will customers find them easy to use together? A distinguishing feature of Apple's music management system—the Mac, the iPod, iTunes, the iTunes Music Store, the FairPlay copy protection software, and, later, the iPhone and iPad—was that they all worked together seamlessly. One underrated innovation of the first iPod was its

FireWire data transfer protocol. If the iPod had used a traditional USB connection, the process of filling its disk with music could have taken up to a full day. By adding the much faster FireWire, Apple ensured that iTunes connected much more efficiently to the iPod.

Connecting innovations can involve everything from the simple matching of shapes and colors, to the physical mating of two devices, or to a complex electronic interconnection. It can also, and often does, involve purely conceptual connections. Think of the links that tied together the many steps that CarMax took to create a stress-free buying environment: no-haggle pricing and the physical layout of the dealership, to name two. Sometimes this connection is simple and easily specified; other times it's complex and requires significant change in the key product. The simpler and more intuitive the connection, the lower the risk involved.

Control: How Will You Manage the Innovation?

Once you've assessed the ability of a complementary innovation to deliver on your promise and its ability to connect to other components of the portfolio, you'll be able to decide how to control it—that is, how to manage its development.

The degree of control it requires will largely depend on your earlier assessments of its importance to the promise, the difficulty of doing or creating it, and how hard it will be to connect with the key product and other complementary innovations. A new and difficult innovation that is critical to delivering the promise and must be tightly connected with the key product is a much higher risk. It must be managed much more closely than an easy, less important innovation that connects simply with other innovations. An example from USAA, an insurance company, will illustrate this difference.

Since its beginning as a provider of auto insurance to military officers, USAA has expanded into other types of insurance as well as banking, investments, and pension management. Often recognized for its high level of customer satisfaction and financial strength, it has become a master of the Third Way.[2]

Consider the differences between two complementary innovations that USAA has adopted. The first was its expansion into banking, and the second was its use of aerial drones to assess property damage after a tornado or other natural disaster. Drones would seem to be the riskier innovation—companies evaluating the commercial use of drones in 2016 faced unproven technology, an evolving regulatory environment, and had few examples of successful commercial application to learn from.

While using drones may seem the riskier of the two innovations, when examined as a complementary innovation to property and casualty insurance, expanding into banking services represented the greater risk. If integrated well, banking services could play an important role in helping USAA fulfill its promise: to provide all the financial services its customers needed at "life event" moments—getting married, buying a house, retiring, and so on. But any banking venture would need to be integrated seamlessly into all the insurance products that USAA offered—a complex and difficult process—and a botched integration would put USAA's promise at risk.

That complexity of connection, coupled with the importance to the promise, is what makes banking the riskier complement for USAA. Adding banking required an intensive redesign of data standards, the addition of new software functionality, the restructuring how financial products and services were sold, and a reorganizing of customer service operations throughout the company, just to mention a few of the challenges. This redesign took USAA years and cost over $1 billion.[3] Yet at the end, the company was able to offer its customers a more comprehensive, seamless, easy-to-navigate, and useful set of financial products.

Contrast that with using drones to inspect damaged properties rather than sending a live inspector on site. When a tornado or flood hits a neighborhood, insurance claims adjusters must fight through debris and other chaos to assess the degree of damage to such insured properties as a house, car, or boat. USAA is replacing that insurance adjuster in a car with a remote pilot-operated, unmanned drone that can take high-definition pictures. The result is a much faster assessment of damage and payment of claims. Speed makes a big difference—getting a check into a customer's hands just a few hours faster allows the customer to be the first to line up a contractor, which can make a difference of months in how long the customer must wait to get his or her home repaired.

Acquiring a bank, a complementary innovation for USAA, was a higher risk because with each complementary innovation, you need to assess *the risk to the delivery of the promise*, not the risk of the innovation itself. Offering banking services was a natural extension for an insurance company. But the difficulty of connecting it to USAA's key product—insurance—and other complementary innovations put the key product and other innovations at risk.

In spite of the novelty of drones, evaluating them according to their risk to the delivery of the promise leads to the surprising conclusion that they're not a risky complementary innovation. USAA was able to test different types of drones, assess the capability of the companies offering them, and wait until it found an acceptable partner. The use of drones never put the delivery of USAA's promise at risk. If they didn't work, USAA could simply go back to the old way of settling a claim.

It's not unusual at this stage to discover that you lack the corporate knowledge and skills to produce the complementary innovation. Your choices then will be to develop those skills internally, find one or more outside partners, or buy an outside organization with the necessary skills and experience and bring it in-house. Each option carries its own risks.

Partnering with an outside company can work, but it requires a set of management skills many organizations lack. This option isn't a simple matter of hiring outside skills such as legal counsel or auditing. Partnering with an outside organization to produce an innovation that must be integrated with other innovations is a much more complex arrangement.

Most organizations have never had to develop the skills and experience to make that arrangement work. There are two general types of partners to choose from: companies that make money selling a particular product or service you need, and consultants that make money selling knowledge about a particular product or service. Consultants are often more expensive in the short term, but can help you develop in-house expertise. External partners are often less expensive and make it easier to change solutions in the future as your needs change.

Bringing the skills in-house via acquisition carries a separate set of risks. How long will it take for the new group to be integrated into the larger organization? How good a fit will there be between the new group and the organization? These are critical considerations in the Third Way, which usually calls for a high level of integration among all the complementary innovations. It's less likely to succeed if an important group operates off by itself.

These are genuine risks even though they're hard to quantify or assess concretely. If you're comparing two competing and mutually exclusive innovations and one of them requires an outside partner, that's a key consideration. Much of the risk will depend on your organization's management flexibility and its experience in coping well with new and different relationships.

We recommend plotting a potential complementary innovation on the matrix shown in figure 6-1. Positioning an innovation on the matrix will tell you what type of control structure will be best. Expanding into banking for USAA would go in the upper right corner because it requires a great deal of adjustment in the rest of the business, and is important for the delivery of the promise. USAA, unsurprisingly, acquired a bank for this expansion and integrated its operations into the rest of the company. Using drones to replace insurance adjusters is on the middle left—the integration is simple and the overall impact is more limited. USAA surveyed the partners available and chose the outside

FIGURE 6-1

Complementary innovation matrix for USAA insurance company

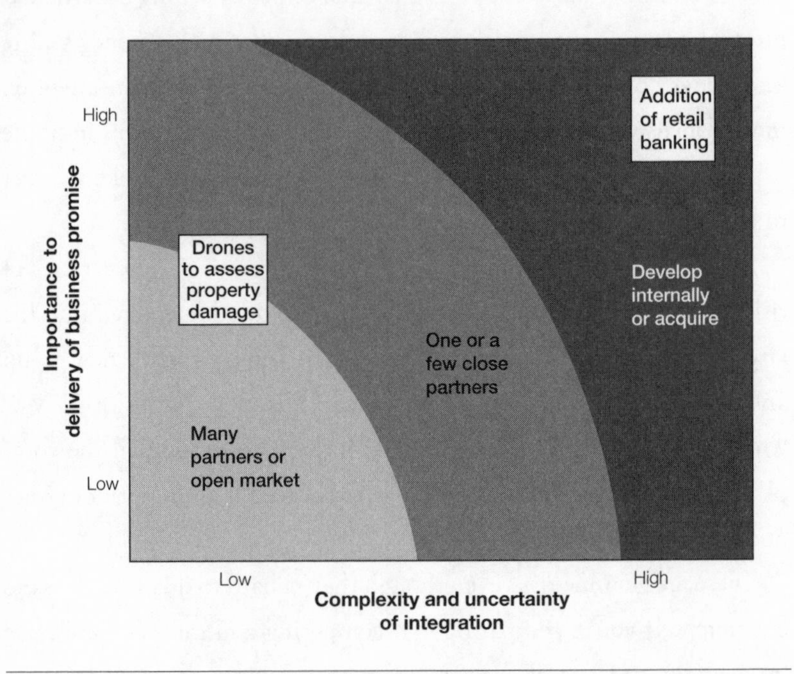

company Precision Hawk as its drone partner. Its first test of the drones happened in April 2016, when a destructive hailstorm hit the company's hometown of San Antonio.

Complete: Build the Minimum Viable Portfolio of Complements

Companies that succeed at the Third Way, we've found, tend to start small. Instead of beginning with as many complementary innovations as possible, they assemble what we call the minimum viable portfolio of complements (MVPC)—the smallest set of complementary innovations that will deliver the promise.

Finding the right set of complements is a "Goldilocks" problem––you don't want too few nor too many. Starting with more innovations than the minimum can complicate the effort, soak up valuable resources, and make analysis of results more difficult. Starting with less than an MVPC will lead to failure, not because your efforts were misguided but because you won't have provided the minimum that fulfills the promise. You can grow the system over time, but you must begin with at least a minimum system. How well would CarMax have done if it had launched its first megastore without the availability of financing for car buyers? How well would the iPod have succeeded without iTunes? How successful would GoPro be without its mounts and video-editing software? Your goal is to make sure you have those few key complementary innovations in place and working well together.

If you discover that the complementary innovations you've selected so far do not constitute an MVPC, you will need to back up and work through the process again; you will need to generate new ideas for innovations and select some different alternatives for testing. This time, though, you will have a clear idea of what you're seeking.

Use Experiments to Decide Which Innovations to Deliver

You've generated ideas for complementary innovations and selected the portfolio of innovations you believe will best deliver your promise. At this stage, companies often jump straight to the development and launch of that portfolio. The urge to move forward fast can be almost irresistible at this stage. All the hard work you've done to identify your key product and business promise, coupled with the extensive field research and brainstorming of new ideas you've done, can make the decision to move forward with the MVPC seem like a no-brainer.

But if there's one thing that the field of innovation research has taught us in the past decade or so, it's this: no one is very good at predicting innovation success. Over and over, smart people at well-respected companies spend millions to introduce products that are complete failures.[4]

Two researchers used the data from the crowdsourcing site Quirky to find the best method for predicting the fate of an innovative idea.[5] Quirky bills itself as a "community-led invention platform." It sources ideas from its community of inventors and then makes and sells the best ideas. The products include office organizers, kitchen gadgets, and fitness accessories. At its peak, Quirky received hundreds of product ideas every week from its community. From those hundreds, it selected the best few to bring to market. Inventors received a share of the gross revenues from their products. Quirky made all the data about its innovation proposals, ratings, selections, and sales available to the public.

Quirky staff—internal experts with deep experience in product development—evaluated each innovation and assigned it a Quirky score, a numeric value that indicated the innovation's likelihood of success. The two researchers also hired outside experts, senior marketing people from Rubbermaid, The Sharper Image, and other well-known consumer products companies. These outside experts averaged twenty-six years of experience in their fields. In addition, the researchers added a third group of prognosticators: prospective customers who were shown the idea and asked, "Will you buy this?"

The best predictor of innovation success, it turned out, was to ask a few consumers if they would buy the proposed product. The outside experts were also pretty good, but the purchase decisions of four random consumers were generally just as accurate in predicting success as were the ratings of seven experienced outside experts. How good was the third metric, the Quirky Score, the one created by the internal Quirky experts whose jobs depended on their ability to choose the best ideas?

The correlation between Quirky Score and idea success was slightly negative. Quirky would have been slightly better off taking the recommendations of its internal experts and doing the opposite.

Note that in this study, prospective customers were *not* asked, "Will this product be a success?" It's unlikely that customers would be any better at predicting the outcome than inside and outside experts. They were better, however, when asked, "Will you buy this?" because they were only being asked what they would do as individual consumers. In predicting their own behavior, they were, unsurprisingly, true experts.

The overall conclusion from this and other research is that managers and other inside experts must become much more aware of, and honest about, what they don't know and more humble about their ability to predict innovation success. While this open attitude is a challenge for any type of innovation, it is especially difficult for companies pursuing the Third Way. Moving away from a key product to complementary innovations means moving away from what a company knows best. Choosing the right innovations is difficult enough when the product domain is familiar; the Third Way calls on leaders to make those crucial choices in new and unknown areas.

Fortunately we've learned a great deal about how to do this phase of the process well. Entire fields of study, such as the lean startup movement, have emerged recently and developed a powerful set of techniques to assess whether an innovation will succeed. As with the other three filters or tests for decision 3, we will quickly outline the steps in this part of the process and provide resources for those who wish to learn more.

How to Predict Innovation Success? Use Pretotypes and Ask, "Will You Buy This?"

The best way to assess whether any innovation is going to be successful is to create a series of quick, cheap tests that accurately capture how real customers buy and use the actual product. These early test versions of a product are known as *pretotypes*.[6] Most people are familiar with a prototype, a preliminary working version of a new product. Prototypes are one-off working versions of an ultimate product and are usually created at the end of the development process when a decision to proceed has already been made. Their purpose is to iron out final issues with the product or service. A pretotype, on the other hand, comes much earlier in the development process, before a decision to proceed has been made. A pretotype is a rough and even crude simulation of the product or service that's good enough for a prospective customer to make a buy-or-don't-buy decision.

Founders of Warby Parker, the online eyeglass retailer, used paper copies of PowerPoint slides to simulate its website design before ever hiring a coder or building an actual website.[7] The company found it could simulate the user experience, get feedback, and fix problems cheaply and quickly using paper. Such pretotypes—often called Frankenstein pretotypes because they're crude and ugly—are good enough to simulate some key component of the final experience, and provide valuable feedback early in the development process.

Innovators have created clever ways to test quickly and cheaply what doesn't yet exist, and they have assigned some vivid and memorable labels for the different types of tests.[8] Here is a small sample:

In the *Wizard of Oz*, something looks real from the outside, but inside is a live person who does what ultimately a piece of software or a machine will do. For example, the maker of the Cointar machine that converts coins to paper money wanted to expand by offering iTunes and other gift cards. For the test, it put a person inside the machine who provided a card when a customer took that option. It was a way to test the idea without having to redesign the machine and build an expensive prototype.[9]

With the *fake-door* test, you offer a product or service that doesn't exist to see how many people express interest. If enough do, you produce the product. This approach is sometimes used in executive education, for example, where a firm or school might say, "We're planning to offer this program on the use of big data in consumer product marketing. Sign up now." In fact, there is nothing actually scheduled, though the program has been designed sufficiently to describe its contents and benefits. If enough people sign up, it will actually be designed and offered; if not, those few who did try to sign up will be notified. Anyone using this approach must be careful with language and not guarantee that the program will actually be scheduled on a specific date. For a product, the language might be "Will you buy this product, when it's available?" rather than a promise for a confirmed delivery.

In the *one-night stand*, you actually do sell the product or service, but only in one place for a short time to one audience. Buyers actually get a product, typically one that's been handmade or adapted for this purpose. If it's a service and the test outcome is negative, the service is still provided for the few who bought it. Producing the product or service is often uneconomic when done at such a small scale, but the cost here is minor. Pop-up stores are a popular version of this test, though they may last more than one day.

There are other approaches—*imposter*, *Frankenstein*, *Pinocchio*, and more—that are easy to find online. All of them aim to generate early, inexpensive, and low-risk information about market interest or customer usage behavior with a new product. Though they differ in how they work, they are all intended to convert assumptions into facts. And all of them, to varying degrees, are meant to pose the critical question to prospective customers: not "*Would* you buy this?" but "*Will* you buy this?"

The Importance of Keeping an Open Mind

As you do these tests, you will inevitably be surprised. Ideas that you thought would be sure successes turn out to have fatal flaws. It also works the other way. You can be equally wrong about something you were sure would fail. So be open to testing that interesting but slightly crazy idea your young intern proposed. You might be pleasantly surprised by the outcome.

The story of Valve showed more than the potential profitability of the Third Way. It also illustrated the evolution of a company pursuing the Third Way. Valve frequently conducts the kinds of tests described above. In addition, it also created a powerful set of tools that allow its partners—game developers—to test their own game ideas. Greenlight, for example, allows game developers to quickly test and get feedback on different game concepts, challenges, graphic styles, and much more before committing major resources to actual development. Valve's Early Access program then lets the developer test a new game, get feedback, and rapidly improve it. By providing these tools, Valve ensures a steady stream of great games to its target customer—the passionate gamer.

Decision 3 can seem at first glance to be a collection of well-known activities that most companies have already mastered. But, as we said in the beginning, occasional twists in the process can trip up a company that's not careful. Innovations that seem low risk can actually pose a major risk to your promise, all the more reason to conduct tests quickly and well. Many companies have mastered these challenges, but no company can take them for granted.

Making Decision 3 Work: Management's New Role

As a Third Way project progresses, the new types of activities that the teams will be doing will require new types of supervision. Many organizations struggle to pursue the Third Way because it calls on them to change the way they manage innovation, and managing a new activity in an old way is the surest way to derail it. Of the three decisions thus far, decision 3 is where this struggle becomes most apparent. Perhaps the best way to understand this is by recalling the two big ideas that this decision calls on managers to embrace.

First: the best way to generate ideas for complementary innovations is by examining the whole context in which customers realize a need for, buy, and use the key product. A crucial part of this examination is to understand what the customer is trying to accomplish. For many organizations, this approach requires a deeper and more empathetic understanding of the customer than is called for by simply focusing on the product and how to improve it.

Second: to choose the innovations they will pursue, organizations should rely on tests and experiments that ask prospective customers, "Will you buy this?" No other way of predicting outcomes is better—neither the opinions of experts nor the judgment of managers and executives.

The success of any Third Way effort depends on accepting these ideas, but the way most organizations innovate doesn't reflect them. Those charged with innovation rarely invest the effort needed to understand the entire customer context. They're expected to approach innovation as they would any other goal-focused process in which they're given a target and constraints—for example, a budget—and then expected to create and follow a plan with milestones for achieving that goal. Focusing on the context and looking beyond the product to the underlying job the customer wants done can feel like a lack of focus on what counts, the product.

To determine which innovations to pursue, many firms follow a highly detailed stage-gate process. The stages are sets of activities that innovators work through. The gates are checkpoints where managers gather to hear innovators report on their progress through the stages.

When the innovators have completed all prescribed work in a stage, they receive management approval to proceed to the next.

Unfortunately, this process is better for tracking activity than fostering innovation. The stages are typically overspecified, and the same steps are assumed to apply to every innovation challenge. Innovators report their progress using mandatory PowerPoint templates, and permission to proceed depends on having mechanically checked off the required tasks.

Rather than tracking activities, managers should track progress by what has been learned, especially about prospective customers and the context in which they use the product. Managers should expect innovators to develop pretotypes starting early in the process and put them in front of prospective customers throughout the process.

Managers also need to adjust their role when leading the team through an iterative test-driven process. Such a process requires no less discipline and structure than does a traditional stage-gate process, but the form of that discipline is much different. In our experience, there are four critical transitions that managers have the most difficulty making:

> *Demand clarity about the goals and process for each experiment.* Demand that everyone is clear on why an experiment is being done before it's actually done. And make sure that the experiments cover all aspects of the customer purchase and use experience. Besides assessing customers' willingness to buy, innovators might also test price levels, differences in preferences between customer segments, manufacturing capability, technological feasibility, compatibility with the key product, and so on. Expect clarity in each test about what is being tested and how the results will be assessed. When the test is completed, review and record the results and, above all, what was learned.

Limit the funds committed and invest progressively as results warrant. The whole point of this approach is to limit costs and losses. Many tests will not work, even those that were expected to succeed, and so it's important that managers limit the funds available to ensure that teams test ideas early and cheaply, even when they "know" something will work.

If the tests raise basic questions about the promise or key product, have the courage to face them. The tests are intended to identify the innovations that will fulfill the promise selected for the key product. Consistently weak results, however, can raise questions about the wisdom of the promise or even the key product. We've emphasized in each decision so far the possibility of needing to revisit an earlier decision (or decisions), and that need is no less urgent here. As we noted earlier, a significant percentage of all innovations fail or fall short in their initial form and need to be revised significantly to succeed. It's hard to back up and redo an earlier decision, but it's sometimes necessary.

The most difficult transition of all—believe the results. For many managers, making major decisions—to proceed with developing a new product, for example—is the essence of what they do. It's a big part, they think, of what sets them apart from their underlings. So it's not easy for them to agree that they and their colleagues lack any special ability to predict the success of something new.

It's not easy for an organization to select complementary innovations in this way, but the companies that develop the necessary skills tend to emerge from decision 3 with a set of innovations that are ready to be rolled out and likely to succeed.

Decision 3 in Action: The Birth of LEGO Ninjago

For decades prior to the turn of the century, LEGO had developed new construction toys by having teams of smart, experienced designers in Billund, Denmark, prepare proposals for new toys and present them to management. The managers then picked the winners to bring to market and tracked their development using a structured stage-gate process.

When LEGO adopted the Third Way approach to innovation, it realized that its promise, especially to young boys, involved creating rich, compelling stories of heroes battling villains that kids could play out with LEGO brick sets. And these brick sets had to be accompanied by TV shows, books, video games, events at the LEGO store, and other complements to get kids involved with the drama in the stories.

So LEGO changed its development process. Rather than have managers in Denmark choose new toy themes and stories, LEGO now turns to seven-year-olds in Fort Lee, New Jersey. The new process begins when creative designers at headquarters generate not toy mock-ups but evocative sketches of LEGO mini-figures ("minifigs") in the middle of different adventures, such as battling robot sharks underwater or carnivorous plants in a jungle. Then the designers gather groups of youngsters in a room and show them the sketches one at a time. For each picture, they ask the kids to make up a story about what they see happening in the picture. The more stories a picture generates and the more excitement those stories arouse in the kids, the greater the potential of that idea.

One picture that generated a tremendous amount of excitement was a sketch of ninja minifigs battling giant "mech warriors" (human-controlled robots). Kids loved it. But instead of going back to Denmark and creating a new toy based on that setting, LEGO designers created more sketches showing the ninjas in a variety of different environments. In some, ninjas battled lizard people, some of whom rode dragons. In others, the setting was more modern—the bad guys were driving motorcycles and flying helicopters. In still others, ninjas fought mummies swarming out of a crypt.

Ninjago was the name given this LEGO toy-in-the-making. As the designers continued their dialogue with youngsters, they realized they needed to answer a crucial question. If ninjas were the good guys, who were the bad guys?

Again, instead of having some smart manager in Denmark make that decision, the designers created possibilities and asked the kids. Their unequivocal response surprised the designers. (After all, who knows the mind of a seven-year-old better than a seven-year-old?) If ninjas were real historical figures, then the answer was obvious. The bad guys couldn't be lizard people, robots, or monkeys. They had to be skeletons. For the seven-year-old boys, there was no other possible choice.

But what kind of skeleton? The testing continued, with the designers drawing many variations and going back and forth with the youngsters until the designers had developed the concept into a full story with a competitive game. The game was called Spinjitsu, and in it, ninjas battled skeletons and other bad guys while both rode spinning tops. To get the tops right, the designers generated dozens of versions and tested each one. When it was done, the story featured four ninjas, four golden weapons, and a villain, the evil Lord Garmadon, who sought to capture the four golden weapons that would allow him to unleash his evil spell on the world.

To complement the basic Ninjago toy, LEGO commissioned a TV show, now in its seventh season on the Cartoon Network. It released a video game that let kids play out the Ninjago story on their Nintendo or Xbox. Graphic novels told the story, brought kids into the world of Ninjago, generated income, and increased demand for the toy. Plus, events at LEGO stores invited kids to come in and compete with each other in Spinjitsu. The result was a smash-hit toy that would have been impossible to create with LEGO's old way of innovating.

Valve, GoPro, LEGO, and USAA—all masters of the Third Way—not only developed and communicated a clear business promise, they also worked tirelessly to develop a portfolio of complementary innovations to deliver that promise. By constantly creating and testing pretotypes, the companies ensured that their portfolio of innovations delivered on the promise.

At the end of decision 3, you will have completed all the basic preliminaries for putting the Third Way into practice—key product, key customers, the promise, and now the specific complementary innovations that actually fulfill the promise. The final phase, decision 4, includes the steps needed to bring those innovations to market, and is the subject of the next chapter.

Three Takeaways for Chapter 6

- Once you've generated a list of potential complementary innovations, evaluate each according to its ability to help deliver the promise and its riskiness. The best way to assess this is to prepare quick-and-dirty pretotypes and ask prospective customers "Will you buy this?"

- Use your promise to guide the selection of a minimum viable portfolio of complementary innovations. For each innovation in that portfolio, select the internal or external partner you'll use, and choose the partnership arrangement that will allow you to manage each relationship most effectively.

- Supervising managers must learn new ways of imposing structure and discipline on the innovation process. Rigid stage-gate processes will no longer work, and neither will a loose, experimental "lean" process. Managers need to understand how to impose rigor and discipline on a rapidly iterating process, and control funding and resource allocations according to the results.

7

Decision 4

How Will You Deliver Your Innovations?

DECISION 1
What is your key product?

DECISION 2
What is your business promise?

DECISION 3
How will you innovate?

DECISION 4
How will you deliver your innovations?

Thus far, you've chosen your key product; defined a clear, compelling promise; and designed a portfolio of complementary innovations that deliver the promise. Now you must bring those complementary innovations to market by deciding who will do each and how you will deliver the entire project.

In this chapter, we will discuss the challenges raised by decision 4 and how to deal with them. For that purpose, we will focus on the critical role of the manager leading a Third Way project. Get that role right, and many of the other problems will become manageable. Get it wrong, and your Third Way project is probably doomed.

We begin with the story of Guinness, the Irish brewer, and what it did to execute decision 4 well.[1]

Guinness and Irish Pubs

In the early 1990s, managers at Guinness noticed something unexpected: a spike in demand for Guinness beers in diverse markets and communities across Europe. Higher sales were showing up even in places like Germany, France, Italy, and Switzerland, where no one had anticipated much, if any, growth.

When Guinness managers looked into this welcome development, they discovered that much of the growth was coming from areas where successful Irish pubs were located. The growth was coming not only from the pubs themselves, but also from bars and restaurants around the pubs. Clearly, the Irish pubs were introducing local drinkers to Guinness, and those drinkers were asking for it wherever they dined or drank.

At the time, 90 percent of Guinness draft beer was sold in Ireland and the United Kingdom, and so Guinness managers sensed an opportunity to expand their market. But how to repeat those local successes? The company was a brewer, not a retailer. Trying by itself to create a network of Irish pubs around Europe and elsewhere would require a huge, risky investment in a part of the business Guinness knew virtually nothing about.

Wisely, the company's first step was to send a team of designers, restaurateurs, marketers, and real estate people to study the pubs.[2] After looking at some seventy or eighty establishments, the group reported its findings.

Most of the pubs were the genuine article—real Irish pubs that faithfully replicated the great mid- and late-Victorian pubs of Dublin and Belfast. Their owners had made significant investments in the pubs' design. In fact, they had bought the millwork—doors, window casings, cabinets, back bar and mirror, and other signature components—from

Irish firms where artisans plied a unique two-hundred-year-old craft. The pubs also served food, including Irish dishes prepared with fresh, local ingredients.

While most of the pub owners were not Irish, many had family ties to Ireland and most had visited Ireland and studied Irish culture. Above all, they understood how to create the warm, friendly atmosphere that made real Irish pubs uniquely and universally appealing.

Eager to grow the number of Irish pubs in Europe and elsewhere but unprepared to open pubs itself, Guinness chose an approach, which continues today, that grows the number of pubs by helping others open them.[3] First, it identified in Ireland the critical expertise that independent owner-operators would need to create a genuine Irish pub. Second, it created an initiative inside Guinness, dubbed the Irish Pub Concept, that would mentor prospective owner-operators through the process of building, opening, and operating successful pubs.

Each of the design-build firms that Guinness identified in Ireland starts work on a new pub by helping the owner choose the right location for the pub and the pub's layout. Authentic Irish pubs aren't rectangular blocks; they twist and turn so that the customer discovers new rooms around every corner, including a back bar—an important component of any true Irish pub. Then it lets the new pub owner choose which style of pub he or she prefers—country, Celtic, or Victorian, for example—and completes the design specifically for the site where it will be located. Then the firm builds the pub in an Irish warehouse, disassembles it, loads it into a shipping container, and reassembles it on site.

Guinness also identified hospitality consultants, food service equipment companies, real estate professionals, architectural consultants, and site selection experts—in short, a pool of know-how and experience that owner-operators can tap as needed. Finally, it helps the new pub owner find young adults with red hair, freckles, and Irish accents who will go work in the new pub.

It's an informal arrangement. Guinness itself pays nothing to members of the pool; nor are there any contracts linking Guinness and the other firms (although the woodwork in most pubs has "Guinness" carved into it somewhere). The entire venture now exists outside of Diageo Guinness; each firm belongs because it benefits from belonging.

The second major piece of the Irish Pub Concept is the mentoring of prospective pub owner-operators. Because the concept cannot succeed unless the pubs succeed as stand-alone businesses, the Irish Pub Concept helps new owner-operators in several ways.

It works with each of them to ascertain if opening a pub is the right thing for them to do. Part of that determination hinges on their ability to make the substantial investment required. While Guinness and the Irish Pub Concept don't directly involve themselves with financing, they will connect prospective owners with banks and help the owners prepare the plans and projections banks need before making a loan. Once financing is in hand, consultants work step-by-step through the entire development process with each owner—some three thousand steps, in fact—to make sure everything is done in a proper and timely way and to help each owner avoid the mistakes new restaurateurs often make. The consultants also provide training in every aspect of running a successful pub. The training program, appropriately, includes a pub crawl around Dublin.

People from Irish Pub Concept and the design-build firms have traveled throughout Ireland, observing hundreds of pubs to understand how they, though they're all different, create the welcoming, convivial atmosphere that transforms a pub into a genuine Irish pub. Creating this magic, they've found, comes from imbuing each pub with its own quirky authenticity—in the way it's laid out, its architectural details, its millwork, the way it's run, and even in minor but essential touches like real photographs and bric-a-brac from Ireland.

Key to that authenticity is the personal story that the design-build firm helps the owner create. It's the story of how that specific pub, its site, and its owner are somehow linked to Ireland—for example, through the history of Irish immigrants in the area, the owner's family history, or something similar. That story is then told in various media throughout the pub. An image from the story might appear in a stained-glass window, an old photograph, or the millwork or artwork created just for the pub.

In all these ways, each pub is both unique and similar. None is a duplicate of another. Yet each, when done right, captures the spirit of Ireland as a place that is warm, open, and welcoming, a place where complete strangers find themselves talking, a haven from the friction and turmoil of the world outside. Guinness benefits by linking its beer with the appeal of the pub. Those who drink it there return again and again, and they order it in other eating and drinking establishments when they seek to re-create at least a piece of the pub's conviviality.

The Irish Pub Concept has paid off for Guinness. Over a period of six years, some twenty-five hundred Irish pubs opened in Europe, increasing annual sales of Guinness draft beer throughout Europe by about half a million barrels. Where only 10 percent of Guinness draft beer was sold outside Ireland and the United Kingdom before 1990, the share has now increased to 32 percent.

As Donal Ballance, a former senior manager at Guinness who was involved in this venture, said, "If you want to make your brand come alive, the best way you can do it is to build a shrine around it. That's what the Irish Pub Concept did for Guinness." It's hard to think of a better way of describing how the Third Way works. By creating a system of new ventures that worked to the advantage of all players, Guinness achieved the distinctive benefits of this approach: low costs, low risk, and high returns.

The story of Guinness and Irish pubs highlights the benefits and perils of decision 4, where all your prior work comes to fruition. The benefits for Guinness were more customers and higher sales. The perils were perhaps less obvious, but they were real nonetheless. They arose from the fact that executing the Third Way often takes you far away from what your company is good at doing. For Guinness, the challenge was that the complementary innovation it chose—opening Irish pubs throughout Europe—had to be done largely outside the Guinness organization. Owner-operators had to be found, design-build firms identified, and myriad sources of other expertise located.

Guinness wanted to set up and manage that unfamiliar work, though nothing in its long history as a creator, brewer, and marketer of beer had prepared it for such a challenge. Though it understood its end consumers quite well, it possessed little expertise in architecture, financing, construction, hiring, running a successful restaurant business, and the many other skills needed to build and manage an Irish pub.

Problem: Leading a Third Way Project through an Existing Innovation Process

Many leaders try to pursue the Third Way, but their organizations stumble when they get to decision 4. The reason is that most companies have designed their internal roles, metrics, and processes to support and encourage more of the same—that is, incrementally better versions of current products or new variants of current products for new market segments.

Those roles, metrics, and processes are designed to reduce the risk and increase the success rate of such new product development ventures. But will they work when you're developing a complementary innovation that's not more of the same?

Try this thought experiment. Turn back the clock and imagine that you work for Guinness in the early 1990s, and the company has decided to develop the Irish Pub Concept internally using the same system it has used for decades to develop successful new beer varieties. You are the product manager who has developed many of those varieties, and you've just been challenged to grow Guinness sales in Germany by developing a new beer specifically for that market.

You and your group do field research and discover that Guinness sells better in German neighborhoods that have Irish pubs. In fact, you meet individuals in your research who express interest in investing in and owning an Irish pub. You return to Dublin convinced that, to succeed in Germany, Guinness needs both a new beer variety and a program that fosters Irish pub ownership across the country.

You make a presentation to Guinness leaders, and you convince them to go forward with both the new beer and the German pub program. Their approval initiates the company's well-established process for managing innovation, which is essentially a series of stages. At the end of each, you will report on your progress, and if the leaders are happy, they will allow you to proceed to the next stage. This is the tried-and-true product development process the company has used for many years with good results.

Now think about how this project might unfold:

- You, the product manager, are an experienced and respected leader. But you know nothing about building a pub, and so you will have to find and work with outside experts. You've never had to do that before in your role. Even more problematic is that you, in your product manager role, have no authority to reach out and form partnerships with outside architects, builders, interior designers, real estate experts, restaurateurs, and financing companies.

- You are given a team of experienced, dedicated, and smart people to work with you on the pub concept, as well as extra part-time support from the business development, legal, and finance departments. But the team members you're given also know nothing about building and launching a pub or about forming partnerships with outside firms that do know. Many team members, in fact, have never before been asked to innovate in this way. And the supervising management team that you report to lacks the knowledge and experience needed to provide you and your team the support and guidance you need.

- The company's standard product development process requires that you develop a detailed business case, outlining the investments required, the profits expected, and the timeline for the various outlays and income. Your business case is now split into two: one for the beer and one for the pubs. The business case for developing a new beer is straightforward and familiar. You know the investment required, the risks, the marketing costs, and the potential payback. But the business case for the pubs is much more difficult to construct. You've found prospective owners who will invest in the first few pubs, but you know that additional investment will be needed to build a program that encourages more owners to open pubs in the future. You suspect that management is unlikely to invest the company's limited funds in this project at the expense of others with better, faster, and lower-risk returns.

- As you construct a project plan, you realize that the pub concept will take much longer to develop than the new beer. You present your plan to management and argue for an expansion of the plan and more time. But you're still given a very tight time frame to both develop the beer and launch the first pub. That's a problem because growing the number of Irish pubs in Germany will take much longer than developing the beer, and your business case for the success of the beer depends on a boost of sales from your new pubs.

This is, of course, an extreme example. Guinness didn't even think about trying to develop the Pub Concept this way. But we suggest this thought experiment because it illustrates the pitfalls that are harder to see in less extreme settings. It's obvious that Guinness would have failed if it had tried to treat the Pub Concept as if it were a new beer. But this approach, unfortunately, is what many companies try to do. They have one process for innovation, and they put every innovation project through that process. The result for a Third Way initiative is almost certain failure.

The goals of a traditional product development process—to manage risk, reduce the cost of development, and increase the probability of success—are still the right goals. But a portfolio of complementary innovations must be developed differently, by different people, and judged by different standards because it will deliver benefits to the organization in a different, more complex way.

A New Leader for a New Approach

The starting point for any Third Way project is to recognize that the conventional methods and techniques most companies have always used won't work anymore. To succeed, the Third Way calls for a different approach, and a different approach calls for a different kind of leader. A traditional product manager cannot do this job. What's needed is a new leader who has the right mandate, as well as the right tools, for the job.

A New Mandate for the Project Leader

When a company designates a key product, it must also name a project leader who will manage all the innovation efforts around it. Most companies assume––mistakenly––that this person should be the traditional product manager responsible for the product.

To understand the problems this approach can create, look again at the GoPro story, but now look at it from inside a key competitor. When Sony saw GoPro emerge as a serious competitor, its executives challenged product managers in Japan to produce a "better" action camera.[4] The product development team succeeded by producing an action camera that was superior in several dimensions: more pixels, image stabilization, better GPS, and a third lower cost. Unfortunately, it was a case of winning the battle—a better camera—but losing the war. GoPro still sold many more cameras. Why? In spite of a better camera, Sony still lacked good mounts, a good smartphone app, PC software, and a good social media site—all crucial complements that helped customers capture their adventures.[5]

Sony product managers succeeded with the goal they were given. But that goal was based on the traditional definition of a product manager's role—to focus solely on the product itself. In setting that narrow goal, Sony leadership focused on beating the competition rather than "dating" the customer. Sony didn't support what the customer was trying to do with the camera. This is how product managers are set up to fail when the situation calls for the Third Way. If they are limited to the traditional goal of producing "a better product," as the Sony product managers were told to do, they can succeed in achieving the goal while simultaneously losing the market.

The scope of the project leader's authority is crucial. Defining that role too narrowly can kill any chance of Third Way success, because, as we suggested in chapter 3, this approach calls for a higher-level role, a *solution integrator*. The most important responsibility of the person in this role is to create and lead a cross-functional team that is separate from but connected to the hierarchy. The job of the leader and team is to spearhead the cross-organizational process of working through the four decisions: designating a key product; finding the strongest possible promise for that product; selecting, specifying, and designing a complete set of complementary innovations around it; and then choosing and managing, inside and outside the company, those who will deliver the innovations.

Mindset is critical throughout this process. The leader and team need to be guided by a deep understanding of the customer and the context in which the customer uses the product. What is the customer's entire experience of, first, sensing a need for the product and then finding, acquiring, and using the product? What will it take to make all parts of that process work better? What more can be done to help the customer solve the entire problem he or she is trying to solve when buying the product?

The Third Way can only succeed if the solution integrator and team possess the ability, mindset, and mandate to lead both a companywide effort and outside partners as well. And it can succeed only if the leader and team are held responsible not just for developing or improving a product but also for helping the customer get value from the product, even if success requires the leader and team to take steps that reach across the organization.

Does this mean you should eliminate the role of product manager? On the contrary, decisions about allocating product development resources will become even more complex, and there's still a strong need for someone to manage the product and ensure that it has the necessary features and quality. But this role is no longer the final decision maker about the product. The allocation of development resources will be based on the needs of internal and external customers. Partners developing complementary products will need information and support, and the person managing the development of the key product will be as busy as ever.

Table 7-1 summarizes the key differences between the old role of product manager and the new role of solution integrator.

If the Sony product managers had been given the right mandate and authority—to meet the needs of adventurers who wanted to record and share their greatest accomplishments—the outcome might have been much different. Those managers and their teams were probably doomed to failure before they started, by the structure, process, team members, and goals that they were given.

Because it's so vulnerable to such organizational issues, decision 4 differs from the other decisions in that it begins early and continues throughout the project. Decision 4 is a set of interlocking smaller decisions, all of which will begin to get made very early in the project, but many of which will not make their effects known until much later. In other words, bad decisions made early can doom a project later.

TABLE 7-1

The evolution of the Third Way product leader: from product manager to solution integrator

Product manager responsibilities	Solution integrator responsibilities
• Collects and codifies user needs for product • Defines product • Tracks competition • Documents product requirements • Prepares product line strategy • Tracks and helps manage product development project • Prepares for launch of product • Tracks sales, usage, and profits from product	• Collects and codifies user needs for the entire system • Assembles solution • Tracks competitors and potential partners for each component in system • Documents solution requirements • Prepares system development strategy • Tracks all development projects, inside and outside company • Prepares for launch of entire system • Tracks total costs and profits

One of the most important responsibilities of solution integrators and their teams in these early stages will be to prepare management, team members, and, later, outside partners for the new challenges of the Third Way. For this purpose, the solution integrator and his or her team will need to create simple, early versions of the overall project specification, business plan, and project plan. These will be revised and expanded as the project moves into decisions 3 and 4.

The *project specification* defines the product, complementary innovations surrounding it, and expected project outcomes. While many of the documents that make up the project specification are familiar to innovators, we recommend adding a new one—the *innovation matrix*, described below—which will help summarize and communicate the project and its goals.

The *business plan* makes financial projections of income, investments, variable costs, and projected profitability for the project. Most companies have specific methods and formats for computing a project's projected revenues and costs. But as with everything else about a Third Way project, there are a few differences in the way a team should construct its business plan, differences that can make or break a project. We will discuss these differences below.

Finally, the *project plan* lays out who will do what, when, to achieve the expected outcome. This is a more straightforward exercise. While a project plan is never simple, if a project team has done the other steps carefully, the planning should not be different from the planning done for any other large, complex project.

In the sections below, we'll talk about project specifications, business plans, and project plans for Third Way projects. The goal, as in other parts of this book, is not to review existing and well-understood material but to show where a different approach will prove useful.

A New Leadership Tool: The Innovation Matrix

Every innovation project needs a clear statement of tasks and deliverables, and a Third Way project is no exception. With traditional product development, the structure and contents of early deliverables for the product—customer needs, product specifications, product costs and price, and investment targets—are usually well understood. As a company makes the transition to the Third Way, however, it will need to create an additional set of deliverables that cover both the key product and the multiple complementary innovations around it. Many of these deliverables will be straightforward adaptations of existing templates. But two in particular will be different for the Third Way—the overall project plan and the business case. First, the project plan.

In February 2014, LEGO released the company's first full-length animated feature film with the LEGO brand attached. While LEGO had worked with outside groups to create direct-to-video movies and episodic TV shows, it had never put the company's name on a full-length feature film. Another major challenge of the movie was that it was done using an entirely new type of computer-generated animation, in which every scene was made to look as if it were created with real LEGO pieces. This type of animation represented a huge technical challenge—creating waves in a sea of blue LEGO bricks, for example, was no small feat.

To track the different parts of this Third Way project, LEGO would have used the innovation matrix (figure 7-1), a project management tool it developed internally. The innovation matrix lets a team track and coordinate the different components of a project and their associated levels of risk.[6] It's a project management and communication tool used to ensure that every part of the project is coordinated well and completed on time. Here is how LEGO might have used this powerful tool to produce *The LEGO Movie.*

FIGURE 7-1

A sample innovation matrix for *The LEGO Movie*

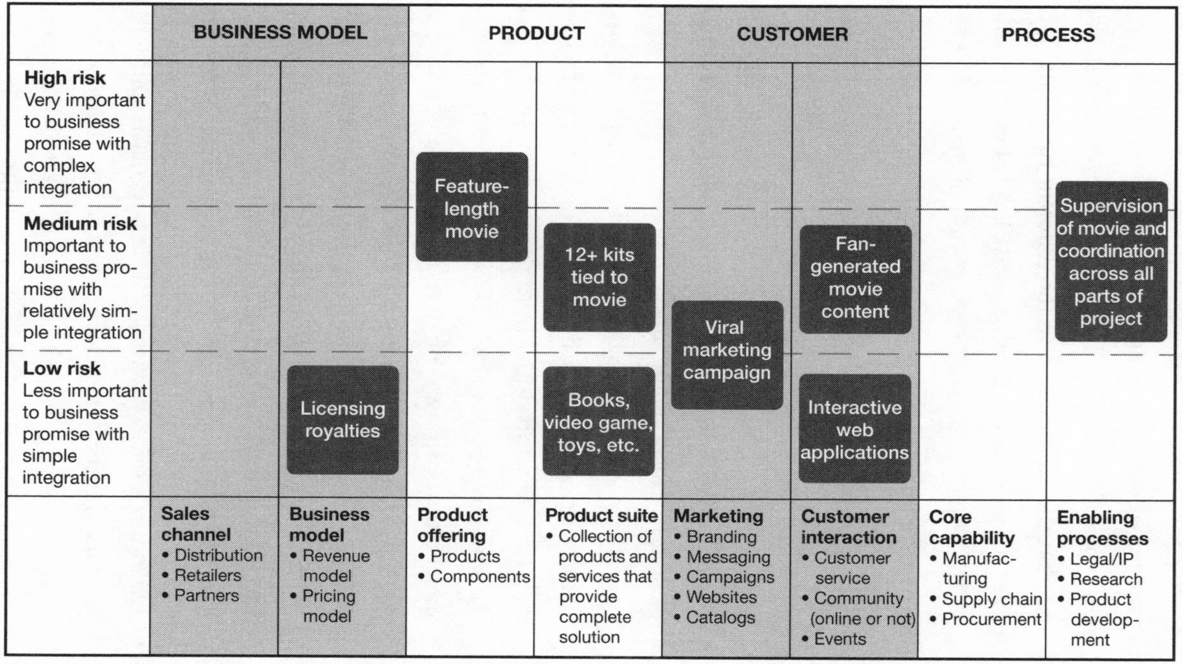

	BUSINESS MODEL		PRODUCT		CUSTOMER		PROCESS	
High risk Very important to business promise with complex integration			Feature-length movie					
Medium risk Important to business promise with relatively simple integration				12+ kits tied to movie	Viral marketing campaign	Fan-generated movie content		Supervision of movie and coordination across all parts of project
Low risk Less important to business promise with simple integration		Licensing royalties		Books, video game, toys, etc.		Interactive web applications		
	Sales channel • Distribution • Retailers • Partners	**Business model** • Revenue model • Pricing model	**Product offering** • Products • Components	**Product suite** • Collection of products and services that provide complete solution	**Marketing** • Branding • Messaging • Campaigns • Websites • Catalogs	**Customer interaction** • Customer service • Community (online or not) • Events	**Core capability** • Manufacturing • Supply chain • Procurement	**Enabling processes** • Legal/IP • Research • Product development

Note: "IP" stands for intellectual property

No two companies will create the same matrix. The unique nature of the key product at the center of the project will drive the definition of the categories. But the overall structure of the matrix should be the same: the horizontal axis is a set of business categories that define the different types of innovation. These different categories should be structured so that they correspond to the different functional groups in the company. The vertical axis shows the risk level of each innovation in the project. The result is a one-page description of the total project.

The first step in building a matrix is to define the horizontal axis—the different types of innovation. These types fall into four general categories. Left to right in the matrix shown in figure 7-1, they are as follows:

- *Business model innovations:* These included new revenue models, pricing schemes, and channels to market. LEGO split this into two subcategories: the Sales Channel and Business Model. For *The LEGO Movie,* LEGO generated income by licensing the characters and stories from the movie to producers of clothing, backpacks, alarm clocks, and other types of branded merchandise. This was not a risky or difficult creative challenge for LEGO, and it generates a significant amount of revenue if done well. LEGO might have also explored innovations in the other category, the Sales Channel—for example, by selling LEGO toys in the lobbies of movie theaters. If it had done that, it would have appeared in the far left column of the matrix.

- *Product or service innovations:* This category is split into the key product and the complementary products and services around it. In the case of *The LEGO Movie*, the film itself represented a major challenge and opportunity. While making movies was by no means core for LEGO, the movie would have been the key product for the team managing the movie, and the LEGO kits would have been complementary products. The team would also have worked with outside partners to develop other complementary products, such as PC games and books.

- *Customer innovations:* This group can include marketing innovations, customer experience innovations, and the development of customer communities. LEGO invested a great deal into supporting its communities of adult and adolescent fans, so it separated community development activities from more traditional marketing activities. One marketing innovation that the company's movie team invested in was an agreement with the *Ellen DeGeneres Show* to send its writer Adam Yenser to a LEGO store to play pranks on customers.[7] The segment appeared just before the launch of the movie. To involve its fan community in the development of the movie, LEGO created a contest on its Rebrick fan site, where it challenged fans to create their own animated LEGO movies.[8] Brief clips from the best entries appeared in *The LEGO Movie*.[9] The contest was a low-cost way to generate fan interest and enthusiasm for the movie.
- *Process innovations:* These innovations are often split into categories, depending on whether they're part of the product manufacturing and distribution chain or part of other supporting processes. For LEGO, participation with an outside creative team would have represented a significant challenge to the product development process. Ensuring that the story and characters were compelling and appropriate was a high priority for the company, and working with an outside team from Hollywood was a risky, unfamiliar challenge.

These different innovation categories—the columns in the matrix—will usually correspond to different functional groups inside the organization. Where each specific innovation should be categorized will depend on the organizational group responsible. A comic book might be managed by the core project team if it is to be tightly connected to the stories in the movie, while a coloring book that only contains images from the movie might be managed by the LEGO licensing group.

The vertical axis reflects the riskiness of each innovation. In the previous chapter, we discussed how the riskiness of a complementary innovation should be assessed according to its effect on the key product. That assessment can now be used to determine where on the vertical axis an innovation falls. The lowest level of risk denotes an innovation that requires little integration with the key product and is less important to delivery of the promise. High-risk innovations—those that need a great deal of integration and are very important to delivery of the promise—will occupy much of management's time and attention. For *The LEGO Movie*, the dozen or so kits that were in stores when the movie opened represented a medium level of risk, not because they were difficult for the company to produce but because their success was critical to the overall promise of the movie—to create a compelling story that kids could play out with real LEGO bricks.

To use the innovation matrix for managing the Third Way, start by locating in the matrix the key product you've chosen, based on the innovation category (horizontal axis) and the risk level (vertical axis) where it belongs. Then, in decisions 2 and 3, as you identify possible complementary innovations, place each innovation in the appropriate spot in the matrix.

Once you've filled it in, the innovation matrix serves four overall purposes:

- To help the solution integrator communicate to the organization which innovations the team is pursuing.

- To provide guidance on which parts of the organization need to be involved.
- To support the management committee as it tracks and monitors the ongoing execution of the project.
- To raise important questions. Does the team have the skills it needs to manage each of these innovations? With whom will the company partner if it doesn't have the skills? What is the overall risk level of the project?

In short, the matrix is a concise way for the solution integrator to communicate the scope and challenge of the project to everyone involved. And it's an excellent tool to help assess whether the project team has the right skills and experience.

A Third Way Business Plan: Local Investments and Global Profits

Any corporate effort that crosses organizational boundaries will inevitably raise thorny questions of income and expense. These issues are difficult to resolve with outside partners, but can become even more difficult when they cross lines within an organization. For example, transfer-pricing disputes between divisions can be some of the most contentious issues in any company. The success of the effort will depend on explicitly raising and answering these types of questions as quickly as possible.

The first step in finding answers is for you and your team to combine all your Third Way efforts around a key product into a *single business plan* that includes all those efforts, identifies the roles played by each, and shows how those elements combine to produce an overall profit.

The need for an overall profit does not mean every complementary innovation must show a profit. Some innovations, such as new ways of marketing and advertising the product, will be natural expense items that no one expects will directly generate revenue. In that kind of situation, the sticky issue will be deciding whose budget must absorb the expense.

Even trickier discussions can occur around innovations that do produce revenue. For those, the Third Way team will need to decide whether it makes sense to turn a profit on that income. It's an important question because creating a profit for the system overall may mean—for the good of the whole portfolio—that some complementary innovations will not be profitable. Then the question becomes, who will absorb the loss?

Consider just three negotiations that the matrix might have sparked between LEGO and the movie team at Warner Brothers when the two groups were cooperating on *The LEGO Movie*.

- *How much should LEGO invest in the movie?* For *The LEGO Movie*, the author's analysis of The LEGO Group's financial returns in 2014 indicates that the company neither invested much in the movie nor shared much in the movie's profits. The LEGO business model was, and still is, to sell boxes of plastic bricks, and the company was quite successful in selling the toys associated with the movie. But its decision meant that LEGO missed out on the windfall profits produced by the unexpectedly large success of the movie in theaters and after.

- *Who pays for marketing?* There was a very expensive promotional campaign around the movie. But this promotion benefited both the movie and the brick sets that accompanied the movie. Who should pay for that promotion? Our sources tell us that Warner Brothers paid this cost, a standard arrangement for partnerships like this.

- *Who gets the licensing revenue? The LEGO Movie* project spun off more than LEGO toys. Many other types of branded merchandise were available, each of which returned a share of the revenue to the company that owned the rights to the characters. In the case of *The LEGO Movie*, we believe that the LEGO Group received most of that income.[10]

As you construct the business case for your portfolio of innovations, you will inevitably come across challenges like these. If you address them early, as partnerships are being formed, they can be addressed amicably and easily, leaving enough profit for all parties.

The key for constructing your business case is to set *local investment budgets* and *global profit targets*. Investments should be managed very carefully for each individual complementary innovation, but profits must be considered as a whole so that the overall benefit to the company can be optimized. This dual approach may sound simple and obvious, but it can be difficult to achieve. Companies often have rigid cost accounting systems that require tightly defined business cases. The result can be a divergence of objectives, resulting in subteams that strive to optimize their individual profit pictures at the expense of the whole.

For example, suppose the outside team producing *The LEGO Movie* and the toy development team inside LEGO were operating under separate business plans, each separately striving to maximize its profit and minimize investment. Each would have different views about when the movie should open. Although it was produced to appeal to all ages, the movie was primarily created for the core LEGO demographic of five- to nine-year-old kids. Movies like that are traditionally brought out either over the end-of-year holidays or during the summer months to maximize their revenues. But the team producing LEGO toys didn't want the movie to come out near the Christmas holidays. The company didn't need to boost sales of bricks sets during that period, when the challenge was more often keeping shelves stocked. If the business plans had been done separately, the movie team would have released the movie near Christmas time, a move that would have ultimately led to lower sales of the associated toy sets. But the LEGO team had anticipated this issue and retained overall control of the release date. After some difficult discussions between the teams, *The LEGO Movie* came out in February 2014. The LEGO team chose that date because it was quite happy to sacrifice movie revenues to maximize revenues from the toys.[11] This was obviously a difficult discussion with Warner Brothers, but one that LEGO had prepared for from the beginning.

The need for strong and clear management here can't be overestimated. If you're not clear about who or what will or won't make money and how different cost and revenue streams will be allocated, the whole interdependent system you're trying to assemble can fall apart. If every group acts independently to drive hard bargains with outside partners, the result may be an overall reduction in the profits from the portfolio as a whole.

Leading a Third Way Project

Decision 4 is not one discrete decision, but a set of related decisions that begin right at the start of a Third Way project. We separate them into their own category because they're uniquely different. They're concerned with how the project will get done rather than what the team will produce. One of the first of these smaller decisions will be to fill the solution integrator role and give that person the scope of authority he or she needs, as described above. The solution integrator will then begin to build the project team. One critical member will be the product manager, the person who leads development of the next version of the key product. While this person should no longer be the overall project leader with P&L responsibility or final say over product specifications, there still is a need for someone to serve as guardian of the key product.

As the project moves forward into decision 2 (define the business promise), the team will begin the process of understanding both the customer and the context in which the customer uses the key product. To support this step, the team will need to add members who have the experience and skills needed to do in-depth field research and uncover latent needs. Ensuring that this information is captured, documented, and communicated is an important part of the solution integrator's role.

When the project moves to decision 3, choosing specific complementary innovations, the team must expand to include the people who will create and manage the complementary innovations that the team ultimately chooses. As the project team evaluates possible complementary innovations, it may discover that some are in categories that have no representation on the team. New team members will need to be identified from different parts of the company or from outside consulting firms. The goal will be to build a team that has the capabilities needed to select, design, and integrate a complete set of complementary products, services, and business models around the key product. The innovation matrix will help the solution integrator communicate the need for those partners to management.

After choosing complementary innovations in decision 3, the team must then choose who will be responsible for developing each. The key factor in this choice is risk. For whom will creating the innovation be less risky? Using an outside group that has experience in successfully producing the types of innovation you're now considering will usually be less risky than giving the task to an inside group with no relevant experience. But factors other than experience—the complexity of integrating work done outside, for example—also matter because there are risks other than financial. When Guinness chose an outside company to pursue the pub concept, the brewer had to work closely with that firm to make sure the inclusion of Guinness product—how it's stored, served, and represented (in wood carvings, for example), was done well.

As we mentioned above, if you can define the leadership role well, specify the project scope properly, and build a global business case, then the project planning should be fairly straightforward. But that doesn't mean it will be easy or familiar. For example, when LEGO worked with Warner Brothers on the timing of *The LEGO Movie*, it quickly learned that the movie would take many years longer to develop than the brick sets. While LEGO was very familiar with the challenge of tying its toys to movie themes, such as Star Wars and Harry Potter, the company found itself in the unfamiliar role of helping to create the characters and storylines in the movie from the beginning. The result was a project that was much longer, riskier, and ultimately more rewarding than a traditional LEGO play theme.

Senior Management's Role in Third Way Projects

Senior management has a key role to play in Third Way success. Too often, corporate leaders say the right words about the need for an expanded project management role, but then they put product managers in the same old box of expectations and constraints. That is, they still reward them for the quality of their products, rather than their ability to mobilize the entire organization to create a complete portfolio of complementary innovations. If they want something new and different from people hired and trained to develop products, corporate leaders need to hire differently, train differently, reward differently, and set different constraints, targets, and metrics.

Probably the most important task for top management is to set up the Third Way team for success. The key is to set the right expectations for the head of the team, the solution integrator. They should make clear what they expect from this person and the team:

- They should expect not just a great product but a complete customer solution created by selecting, designing, and integrating a coherent system of products, services, and business models around a key product.
- They should expect to see a profitable portfolio, not necessarily a profitable product.
- They should expect the solution integrator to lead a complex cross-company effort, not just a small internal team of R&D personnel.
- They should expect decisions to be made based on market data and experimentation, not the opinions and hunches of managers, including themselves.

After setting these new expectations, senior managers must then actively support the solution integrator as he or she works across the organization and inevitably calls on functions to accept such unappealing outcomes as forgone revenue and higher expenses and to provide the project with skilled people whose functional bosses would have preferred to use them elsewhere.

One Final Piece of Advice

The Third Way is a different approach to innovation. The overall purpose of this book has been to convince you that it's an alternative every company should consider. And the purpose of this chapter has been to help you understand how to adapt your internal processes, roles, metrics, and structure so that, if you do try the Third Way, your teams will have a reasonable chance of success.

Our final piece of advice is simple: start small and start local. A company pursuing the Third Way for the first time is more likely to succeed if it begins with a reasonably sized effort that it treats as a one-off project. It's important for leaders to be clear about what's being done and what this approach requires of people and managers. But mistakes, false starts, and dead ends are virtually inevitable, and so it's usually better not to begin by announcing a big new initiative and making permanent structural changes right away.

There's a significant corporate learning curve that must be ascended. So pursue the Third Way at first as a single project, and focus on doing it well. Give people temporary roles and assignments. Set up ad hoc teams and committees. Then, after gaining experience and getting the process to work reasonably well, institutionalize the approach. The advantage is that when you do make permanent structural changes—new positions, different roles, new teams, new assignments—people will understand what's happening and the chances of longer-term success will rise significantly.

Three Takeaways for Chapter 7

- The Third Way, a new and different approach, calls for a new kind of leader to manage a new process. Putting a product manager with a narrowly defined set of responsibilities in charge will doom a Third Way project. The leader must be a solution integrator who possesses the skills and authority needed to work across the organization with people and groups from multiple functions.

- One of the first tasks of the solution integrator should be to construct an innovation matrix that shows all the different complementary innovations needed to deliver on the business promise.

- Just as the solution integrator role should have responsibility for the entire project, the business plan should cover all phases of the project. While the different parts of the project will have tightly defined and carefully managed investment targets, the profits should be shared and maximized across the entire project.

8

Lessons from an American Icon

We close *The Power of Little Ideas* with a tale both inspiring and cautionary, the story of The Walt Disney Company.

It's inspiring because this American icon, a pioneer of the Third Way, owes much of its success—and, at crucial times, its very survival—to complementary innovation. It's cautionary because the history of Disney is hardly a story of inexorable growth and success. The company's fortunes have waxed and waned in the ninety-some years since its founding, and from that mixed record, we believe there are lessons to be learned about the Third Way.

Anyone looking at The Walt Disney Company today is most likely to say it began with animated cartoons, moved on to animated feature films, which it invented, and then steadily expanded over the years to become a diversified entertainment powerhouse that incorporates animated films, live-action movies, theme parks, television programs, a chain of Disney retail stores, Disney merchandise, and a host of lesser ventures.

We see a different story. Yes, Disney is a diversified entertainment company that evolved from its early success in animated films. But we think the Disney-branded portion of the company—excluding the ABC and ESPN television networks and the complementary innovations around them—is best understood as a practitioner of the Third Way, starting with Walt Disney himself. Much of what is typically considered the company's opportunistic diversification over the years is better understood as complementary innovation around its core, which in this case was not an individual product but a type of product—animated feature films.[1]

But Disney is also a lesson in how successful application of the Third Way can lead to internal dysfunction, separation of the different types of innovation, and ultimately—in Disney's case—an erosion of capability in the core. To understand this, you must understand something about Walt Disney and the history of his company. We'll start at the beginning.

The Animator from Missouri

Born in Chicago in 1901, Walt Disney grew up in two Missouri communities: Marceline and Kansas City. An indifferent student, he was remembered as an artist who drew cartoons and other illustrations for his school's student publications. He left high school before graduating and served a year in Europe at the end of World War I as a Red Cross ambulance driver. On his return to Kansas City, he worked as an artist for a small ad agency, where he learned to create animated movie ads that were shown to moviegoers. This was his first contact with movie animation, and it captured him.

In the early 1920s, animated movies were considered a novelty—only good for showing gags and pratfalls to amuse patrons before the real movie began. After an unsuccessful attempt to set up an animation studio in Kansas City, Disney decided to try his luck in Hollywood because his uncle lived there and his brother, Roy, had been sent there to recuperate from tuberculosis. His choice of Hollywood wasn't as obvious as it seems now. The leading animation studios of the time were actually located in New York City.

In Hollywood, he had some initial success with a series of short films that combined animation with live action, and with fully animated films starring a character of his own creation, Oswald the Lucky Rabbit. When he lost the rights to Oswald to his distributor, he invented a new character—Mickey Mouse—whose rights he carefully protected. The first two Mickey Mouse cartoons gained some attention, but Disney's real breakthrough came when he decided to add sound to the third cartoon, *Steamboat Willie*, in 1928. What made this cartoon distinctive was not just the addition of sound—it was the first talkie cartoon—but the way the sound and onscreen action were tightly synchronized. *Steamboat Willie* excited cartoon audiences just as Al Jolson's *The Jazz Singer*, the first talkie feature film, had thrilled moviegoers the year before. Mickey Mouse quickly went on to become a worldwide phenomenon. His spunky optimism seemed to be the tonic a Depression-weary world needed. By one estimate, a million *audiences* (not just individual moviegoers) saw a Mickey Mouse cartoon each year in the early 1930s.[2]

Animated films in the 1930s were enormously complicated to make. Creating the illusion of motion required Disney and his animators to create 24 slightly different drawings for every second of film—1,440 frames per minute.[3] But as difficult as the process was, Disney was obsessed with doing it well. Of his competitors, he once said, "We can lick them all with Quality."[4] As one of his animators said, "Whatever we did had to be better than anybody else could do it, even if you had to animate it nine times, as I once did."[5]

Yet what truly set Disney apart was not the quality of his artwork but the way he thought about animation. To him, the magic of animation was that with it, he could create any world he wanted. Any fantasy could be turned into seeming reality. If Mickey needed music, he could play a cow's ribs as if they were a xylophone. If he needed a ladder, he could turn his own tail into one. Because Disney could see animation's enormous possibilities, he set his sights higher than did other animation studios. He concentrated on telling a story rather than animating a series of silly sight gags, and so he was the first animation head to create a story department. His goal was to create animated characters so real they could generate an emotional response from the audience. While others thought animation simply meant making characters move on the screen, he understood its real meaning. To "animate" meant "to bring to life" in every sense of the word—physical, emotional, and psychological. He wanted to breathe life into his characters so that the audience would care about and identify with them. No one else at that time saw the possibilities he saw or pushed their animators to reach for them.

From Cartoons to Animated Feature Films

Given such ambition and vision, it was almost inevitable that Disney would start to think beyond cartoons and animated shorts. In the early 1930s, he began to consider producing a full-length animated feature film that theaters would offer as an evening's main attraction. He and his brother Roy, the studio's money man, also hoped a longer film would produce better financial returns. Every time the studio got more money for its cartoons, Walt improved the process of animated storytelling in ways that always seemed to raise costs. He saw an animated feature as a way to improve the studio's financial stability and to stay ahead of its competitors.

Walt chose the story of Snow White as his—and the world's—first animated feature. Producing it would take years and stretch to the limit both the studio's finances and the many animators who worked on it. If it had failed, it almost certainly would have pushed the studio over the edge.

But, of course, it didn't fail. A huge success with both audiences and critics when it was released in 1937, *Snow White and the Seven Dwarfs* allowed the company to pay off its loans, begin to build a bigger studio, and hire the hundreds of additional staff that producing more animated features would require. Over the next three decades, the studio under Disney produced a stream of such features, many of which are still considered animation classics, including: *Pinocchio* (1940), *Fantasia* (1940), *Dumbo* (1941), *Bambi* (1942), *Fun and Fancy Free* (1947), *Melody Time* (1948), *The Adventures of Ichabod and Mr. Toad* (1949), *Cinderella* (1950), *Alice in Wonderland* (1951), *Peter Pan* (1953), *Lady and the Tramp* (1955), *Sleeping Beauty* (1959), *One Hundred and One Dalmatians* (1961), and *Mary Poppins* (1964), a hybrid live-action and animated feature that was nominated for a Best Picture Oscar.

While some of these films were critical and financial successes, many were not. *Pinocchio*—Disney's first film after *Snow White*—returned only 44 percent of what it had cost to make and release. *Fantasia, Dumbo,* and *Bambi*—his next three films—produced lackluster returns—especially *Fantasia*, which attempted to animate classical music. Even though Disney lavished special care and effort on it and even created a special sound system for it, *Fantasia* was a disappointment. The studio struggled through the war years as it made films for the government and armed forces. After the war, through the 1950s and early 1960s, it continued to produce animated features that did well enough—though *Sleeping Beauty* in 1959 was a great popular and financial disappointment.

Disney and Complementary Innovations

By the early 1950s, it was clear to Walt and Roy Disney that producing animated feature films was a very risky business. The costs to produce a new film were high, and audience reaction difficult to predict. Disney's animated film business—if separated from the rest of the company—cannot be considered a success on any commercial dimension. In fact, as we will show next, if the studio had depended only on those films and cartoons to support itself, it would have failed. What allowed the studio to survive and ultimately thrive were the complementary innovations the Disney brothers created around those films. To understand this, we must return to the early days of Mickey Mouse cartoons.

Mickey Mouse Clubs and Merchandise

The first complementary innovation appeared just as Mickey Mouse's popularity was building. The manager of a suburban Los Angeles theater invited Walt to a regular matinee meeting the manager held to fill his theater with youngsters on Saturday afternoons: the Mickey Mouse Club. At club meetings, members took the Mickey Mouse pledge, sang a Mickey Mouse song ("Minnie's Yoo Hoo" with a chorus of farm animal sounds), recited the Mickey Mouse creed, and watched Mickey Mouse cartoons. Walt embraced the idea and authorized the manager to expand the club to theaters all across the country. At their peak, the Mickey Mouse Clubs had an estimated one million members spread over eight hundred chapters from coast to coast. A Mickey Mouse comic strip, which quickly spread to twenty-two countries, was equally popular.[6]

And there was Mickey Mouse merchandise. Mickey's picture appeared on Post Toasties cereal boxes, Cartier sold a diamond Mickey

Mouse bracelet, and there was a Mickey Mouse version of almost every article of children's clothing, as well as Mickey Mouse dolls, comic books, candy, watches, and toy trains.[7] In 1934, sales of Disney merchandise (mostly Mickey Mouse items) totaled $70 million worldwide. That same year, Walt commented that he made more money from Mickey's ancillary rights than he made from the mouse's cartoons.[8]

Mickey was only the beginning. The studio vigorously pursued a similar approach with all its animated characters—for example, in 1938, fans bought $2 million worth of *Snow White* handkerchiefs.[9] By 1947, sales of Disney-related merchandise rose to roughly $100 million per year. By 1948, five million Mickey Mouse watches had been sold, and there were more than two thousand Disney-related products. Sales of that merchandise, in a giant virtuous circle, built the audience for Mickey Mouse cartoons and—later—the company's animated features.

Disneyland

Walt Disney's concept for building Disneyland, which opened in 1955 in Anaheim, California, apparently emerged from the confluence of several of his interests and concerns. Animated features had proven themselves financially problematic, and he tired of the unending financial strains that he could no longer pass off to his brother. Also, and not least, a bitter animators' strike in 1941 had left him feeling betrayed by his "boys," many of whom had worked closely with him to produce *Snow White* and other classics. Whatever the reason, after World War II, his heart seemed to have moved on from animated features and was seeking something else to embrace. In Disneyland, he found the next outlet for his creative energies.

In the 1940s and 1950s, US amusement parks were seedy, dirty, disreputable places that no one would visit with children. After seeing the Danish Tivoli Gardens and American attractions such as Colonial

Williamsburg and Greenfield Village, Walt decided to change that. Just as he used animated features to create a fantasy world for himself and his audience, he would use a Disneyland park to create a physical fantasy world that the "audience" could actually enter and explore. Indeed, core parts of that physical space would be the worlds he had first created on screen. He even thought of the park as a giant movie set that guests could wander through. He insisted that park employees consider themselves actors on stage playing a role when they were working. He planned to surround the park with a berm that blocked any sight of the world outside. Once you entered, nothing would disturb the fantasy. It would be a place and a world never before seen.

It's easy to believe that Disneyland was a natural extension of Disney's animated films but that, otherwise, they were unrelated. We believe the relationship is tighter and more organic than that. Disneyland complements the animated films and wouldn't exist without them. It's impossible to imagine Disneyland without the fantasy worlds and characters Disney created in his films. When Disneyland opened in 1955, *McCall's* magazine called it a place where "Walt Disney's cartoon world materializes bigger than life and twice as real."[10]

In return, Disneyland expanded and extended enormously the economic value of Disney's films and characters. Disneyland put Walt's company, for the first time in its history, on a solid, stable financial foundation. And the creation of Disneyland enabled Disney to develop another complementary innovation: a weekly television program.

Walt Disney and Television

While Disney would not launch his own TV show until the mid-1950s, he had been aware of television's potential since the mid-1930s, when he saw an early demonstration. He considered it a medium he could use and thereafter always insisted on retaining television rights for everything his studio produced.[11]

Disney saw in television a means not only to promote Disneyland, but to finance its construction as well. He shrewdly approached ABC, the newest and weakest of the new national networks (NBC and CBS were the other two), which was desperate for programming to attract an audience and expand the number of affiliated stations. It especially wanted to tap into Hollywood movies, particularly those aimed at a growing market segment of "youthful families."[12]

The ABC-Disney agreement called for the Disney studio to produce a weekly program hosted by Walt that drew on the studio's vast library of films and cartoons, included some new material, and—above all, in Walt's mind—devoted at least one segment per week to promoting Disneyland, which was also the name of the program. In return, ABC would pay for the show's production and invest in the construction of Disneyland. *Disneyland*, the program, launched in 1954 (the year before the park opened) and was such a hit with the national audience that it transformed ABC into a bona fide network. Based on that success, ABC added a daily hour-long program, the *Mickey Mouse Club*, also a great and long-lived success. Together, those programs made Walt a media star.

Disney and the Third Way

It should be clear by now that Walt, aided by Roy, was a master of the Third Way. According to his biographer, Walt was "the first to bundle television programs, feature animation, live-action films, documentaries, theme parks, music, books, comics, character merchandise, and educational films under one corporate shingle."[13] The income from those complements allowed the studio to survive its many brushes with bankruptcy before the mid-1950s and put it on a firm foundation with steady, healthy profits.

Some may disagree. Didn't at least some of those complementary innovations—theme parks, in particular—make the company a different company? Instead of complementary innovations, weren't these efforts just natural business and product extensions, lateral innovations, that allowed the company to evolve from an animated film company to ultimately a diversified entertainment and media enterprise with multiple key products? Hadn't animated features, once important, become just one of many products, and not even the most important, as the company continued to move on and evolve?

This argument may sound plausible, but it wasn't the way Disney himself saw his company. In 1957, two years after Disneyland had opened and its immense success was obvious, he prepared a diagram that shows how he saw his creations: not as a collection of disparate, independent parts but as one organism with a heart that made everything else possible (figure 8-1).

At the center of the diagram are Disney theatrical films, animated and live-action. Around the films are the complementary innovations—Disneyland (theme parks), television programs, merchandise, music, comic strips, publications, and others—based on the films. Note the arrows that indicate both the connection and direction of influence. Films lead to everything else. Films make everything else possible. But note too that many of the complementary innovations also feed and support each other.

Disney's diagram is the best description of a Third Way system we've seen. Take a moment to review the three characteristics of the Third Way: First, it consists of multiple, diverse complementary innovations around a key product or service that make the key product (and each other) more appealing and competitive. Second, the complements operate together as a system or family to carry out a single strategy or purpose—in the case of Disney, to enrich the customer's experience of

FIGURE 8-1

Walt Disney's view of his company

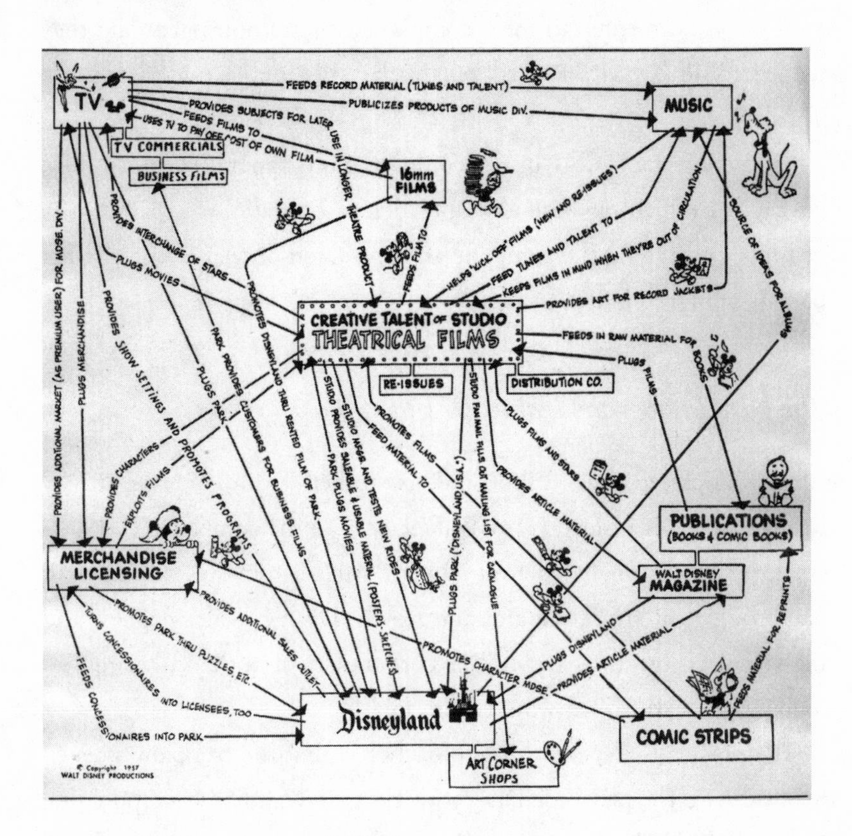

Source: The Walt Disney Company. © 1957 Disney.

the story and characters first introduced in feature films. And finally, the family of complements is closely and centrally managed. Walt clearly designed his company to deliver on all three of these characteristics.

If the story ended here, it would still be an inspiring example of the Third Way. But at the beginning of this chapter, we promised a tale both inspiring and cautionary. This is where—after Walt and Roy died—the story of The Walt Disney Company turns cautionary—and instructive.

After Walt and Roy passed away, those who led the company saw it in different ways. Some saw it as Walt had seen it: a company with a core surrounded by multiple complementary innovations that both supported and fed off the core and each other. Other leaders, however, saw it as an entertainment company with multiple parts, and they tended to deal with the parts separately. They didn't understand the Third Way.

By tracing the history of the company from about 1970 until now, we can see which of these two approaches—pursuing the Third Way or simply managing the company as a collection of related but distinct parts—produced better results.

The New Leadership Team: 1971 to 1984

After Walt passed away in 1966 and his brother Roy in 1971, the Disney board named an insider, Card Walker, to lead the company. The leadership team also included Ron Miller, Walt's son-in-law, whom Walt had brought into the company, and Roy E. Disney, Walt's nephew and the son of his brother-partner. The corporate mantra they all sought to follow was clear: "What would Walt do?"[14]

The practical effect of this approach was that the company pressed forward with the plans for Disney World that Walt had been pursuing when he died. The Magic Kingdom was the first segment to open in that Florida complex, in 1971, followed eleven years later by Epcot, the futuristic community that had most fascinated Walt.[15] In 1983, a year after Epcot, the company launched the Disney Channel and opened Tokyo Disneyland Resort.

Through this period, as the company seemed to move forward aggressively, it was drifting away from character- and story-driven animated feature films. Instead, it produced such forgettable live-action movies as *Herbie Rides Again* (1974), *Escape to Witch Mountain* (1975), and *The Cat from Outer Space* (1978). From 1970 through 1984, Disney released a total of only five animated films and two hybrid animated-live-action features. In eight of those years, no animated features of either type were released. And in 1984, the studio released no movies at all.[16]

Almost all of the few animated features Disney did release in this period lacked the distinction and innovation of Disney's earlier films. Even the company's iconic weekly television program, *Disneyland*, which debuted in 1954 and featured Walt himself, struggled through this period and was finally canceled in 1981.

CEO Walker seriously considered shutting down the motion picture arm of the company, which helps to explain the relative dearth of films released in the late 1970s and early 1980s. Recalling Walt's adage that the only worthwhile publicity was word of mouth, which was free, Walker refused to spend money on marketing and advertising movies. If people didn't like or want the kind of movies that Disney was making, then it wouldn't make movies.[17]

Because of all this, financial results declined. Pretax return on assets, a healthy 13.2 percent in 1973, declined almost in half to 6.9 percent in 1983. Markets ignored the company, and its stock price remained basically unchanged from the early 1970s, when Walt Disney World Resort opened, through the mid-1980s.

By that time, Roy E. Disney had resigned as an employee of the company, though he remained on the board. Frustrated, pushed aside, excluded from decision making, and knowing that Walker had labeled him the "idiot nephew," Roy E. left the company in 1977 to protest the direction management was taking. In his letter of resignation, he said the "creative atmosphere for which the Company has so long been famous . . . has . . . become stagnant," and "the Company is no longer sensitive to its creative heritage." Furthermore, "motion pictures and the fund of new ideas they are capable of generating have always been the fountainhead of the Company; but present management continues to make and remake the same kind of motion pictures, with less and less critical and box office success."[18] Ironically, this man, whose father had headed the business side of the studio, became the advocate of its creative side.

The company's drift away from its creative heritage was felt throughout the company. "When I started here in 1978," said Chris Buck, who years later directed the animated hit *Frozen*, "the studio was run by Walt Disney's son-in-law, Ron Miller. Nice guy, but he wasn't a filmmaker and he wasn't an artist."[19] As if to prove the point after Roy E. Disney left, Ron Miller picked Thomas Wilhite, the publicity director, to take over as vice president of creative development for the company. Wilhite had never before produced a motion picture. His live-action films through the early 1980s did poorly at the box office.

What for a time hid the poor performance of Disney's feature films was what later management would call the "moat," the ring of complementary innovations such as theme parks, merchandise, and television that surrounded the moviemaking unit and protected the company from the ups and downs of the movie business. Unfortunately, the moat also distracted company leaders from the importance of maintaining a healthy film business.

Eventually, the company's poor performance caught up with Walker, and he resigned in 1983. He was replaced by Raymond Watson, an architect who had worked for Walt in the early days of planning Epcot. Picking Watson seemed to support another charge that Roy E. Disney had made when he resigned—that the company was "more interested in real estate development than in motion pictures."[20]

Watson's tenure lasted only a year. By early 1984, the company's stock price had dropped from $85 in late 1982 to $58. From a high of $135 million in 1980, the company's net revenue had fallen to $93 million in 1983. The shareholders, led by Roy E. Disney (who was still on the board), were upset, and new management was hired.

A Return to the Core: The Mid-1980s to the Mid-1990s

The year 1984 marked the beginning of a new era. Michael Eisner was named chairman and CEO, and Frank Wells became president and chief operating officer; each reporting directly to the board. To revitalize the feature film business, Eisner hired Jeffrey Katzenberg to head moviemaking, and brought Roy E. Disney back as head of feature animation, reporting to Katzenberg.

Katzenberg and Disney revived the studio's animation's arm, which sprang back to life with a string of animated hits that included *The Little Mermaid* (1989), *Beauty and the Beast* (1991), *Aladdin* (1992), and, most successfully, *The Lion King* (1994). Seeing the potential in these stories for more than an animated movie, Disney created a theatrical arm whose first venture, a musical version of *Beauty and the Beast*, opened on Broadway in 1994, followed by a musical *The Lion King* in 1997. *The Lion King* went on to become only the fourth Broadway show ever to play seven thousand performances. After the turn of the new century, it was followed by others, including *Mary Poppins* (2006), *The Little Mermaid* (2008), and *Aladdin* (2011).

The company also revived the Mickey Mouse Club on television. The show, now titled *The All-New Mickey Mouse Club*, went on to launch the careers of several stars, including Justin Timberlake, Christina Aguilera, and Britney Spears. In addition, Eisner opened the first Disney Stores in 1987 and expanded the chain aggressively.

As the quality of the animated feature films improved, revenues soared across the board. From 1991 to 1997, Disney's entertainment revenues rose from $2.6 billion to $7.0 billion, consumer products sales went from $700 million to $3.8 billion, and theme park and resort revenues climbed from $2.8 billion to $5.0 billion. In that same period, Disney's overall revenue rose from $6.1 billion to $22.5 billion.[21]

More Stagnation: 1997 to 2006

In 1994, following a well-publicized internal struggle, Katzenberg left Disney to start his own studio. Corporate performance metrics continued to look healthy for a few years, but Katzenberg's departure actually marked the end of an era. Once again, the company focused more on independently expanding complementary businesses than on producing the kind of animated features that moviegoers loved and that ultimately drove complementary sales.

By 2006, the studio had gone twelve years without an animated feature film hit. The reason, according to an article in *Wired*, was the film development process at Disney: "Like most movie studios, [Disney] had for decades employed a C-suite of what's somewhat generously known as 'creative executives'—cookie-cutter MBA types who tasked underlings with turning vague premises into magic . . . Somehow, though, as Disney Animation's films became less successful, the executives exerted more power. They made decisions about what movies would be developed—based on market research, tea leaves, their own opinions—and assigned directors and producers to those projects, none of which became hits."[22]

Don Hall, who directed the 2014 animated hit *Big Hero 6*, first joined Disney in 1995. "It was a broken system," he recalled. "I can't pinpoint where we lost our way, but it was affected by the fact that the people in charge weren't necessarily lovers of the art form."[23]

Films weren't the only trouble spot. The number of Disney Stores grew rapidly, and many stores were located in inappropriate and unprofitable locations. The company was struggling to turn around its Euro Disney park, which had flopped upon opening outside Paris in 1992 and continued to have problems. The Disney-MGM Studios Theme Park at Walt Disney World Resort opened and proved to be an awkward alliance. In moves that, in hindsight, can only appear misguided, the company bought both a pro hockey franchise and a pro baseball team to leverage two live-action feature films, *The Mighty Ducks* (1992) and *Angels in the Outfield* (1994). So tenuous was the connection between these teams and the films they were supposed to complement that Disney sold the Anaheim Angels in 2003 and the Anaheim Mighty Ducks in 2005. It was also in this period that Disney began selling cassettes and DVDs of its classic movies, a trove of great value but not one that could be exploited indefinitely, particularly given the studio's inability to keep growing that trove.

More than 100 percent of Disney's revenue growth from 1997 to 2000 came from the growth of its television properties (the ABC network and ESPN, the sports cable channel). Despite the tremendous success of three Disney-Pixar movies (*Toy Story*, *A Bug's Life*, and *Toy Story 2*), revenues in the studio entertainment division actually dropped from $7.0 billion to $6.0 billion between 1997 and 2000. Income from the consumer products division dropped from nearly $900 million in 1997 to $455 million in 2000. By the summer of 2001, all key metrics for the company, including return on equity, return on assets, and return on investment, had declined by more than 50 percent since 1997. Net income had dropped steadily from $2.0 billion in 1997 to $832 million in 2000. The stock price had dropped from $23 to $17.

And as bad as things were in 2000, they only got worse from there. The stock market tanked in 2000, and the terrorist planes brought down the Twin Towers in 2001. Merchandise sales continued to decline. Theme park and resort revenues dropped. The company reported an overall loss of $158 million in 2001, and performance continued to languish over the next few years. ABC wasn't doing well. The acquisition of the Family Channel was a disaster. Disneyland Resort Paris (renamed from Euro Disney) still struggled. And the company began closing its hundreds of unprofitable Disney Stores.

The one bright spot throughout this period, especially for films, was Pixar. Disney had an agreement with Pixar to distribute and cofinance its films.[24] In return, Disney received half the revenues and owned all rights to the stories and characters. Disney benefited enormously throughout this period as Pixar turned out an unprecedented string of animated hits. The first was *Toy Story* in 1995 (the first computer-generated feature film), followed by *A Bug's Life* (1998), *Toy Story 2* (1999), *Monsters, Inc.* (2001), *Finding Nemo* (2003), *The Incredibles* (2004), and *Cars* (2006). All these movies were huge financial and critical successes and all, of course, gold mines of ancillary merchandise, not to mention the Pixar-themed attractions that Disney could add to its theme parks. The contrast between Pixar's successes and Disney's failures couldn't have been more stark.

The profits that Disney collected from its contract with Pixar and the growth of its ABC and ESPN television properties helped mask the decline in the company's Disney-branded core enterprises: movies, theme parks, stores, and merchandise. In those businesses, Disney was wringing as much value as it could from past success, but it was building no foundation for the future. In Eisner's letter to shareholders in Disney's 2002 annual report, he recognized the problems with the company's earnings and stock price. But he promised that the investments recently made in the Disney and ESPN brands—that is, in the complementary innovations around both of them—would create "a protective moat."[25] He predicted that those businesses would grow annually by 20 percent—but made no major changes in the animation business that could have driven the Disney portion of that growth.

It was no small matter, then, when Disney's relationship with Pixar unraveled in the early 2000s. When Pixar proposed *Toy Story 2* as one of the three films it would produce under its existing agreement with Disney, Eisner said no. After that, Steve Jobs, chairman of Pixar, refused to negotiate with Eisner and instead went through others at Disney. Pixar extended the agreement to cover seven films, but ultimately, the relationship between Jobs and Eisner broke down. In January 2004, Pixar announced that it was ending talks to extend the agreement.[26]

Eisner, already under attack by dissident investors, including Roy E. Disney, resigned on September 30, 2005, one year before his contract expired. Bob Iger was named his replacement.

A Return to Animation: 2006 to Now

One of Iger's first moves was to reconnect with Jobs and negotiate Disney's purchase of Pixar in 2006 for $7.4 billion. Iger understood the importance of animated features, an importance impressed on him when, as the new CEO, he traveled to Hong Kong in September 2005 for the opening of Hong Kong Disneyland. As he watched the opening parade, he realized that all the Disney characters he saw on parade floats were based on films decades old, and all the recent characters were from Pixar movies. "It was pretty easy to see that we had a real problem," he said. "It was staring me in the face."[27]

When Disney bought Pixar, it decided not to combine the two animation studios. However, it did move three of Pixar's key leaders, President Ed Catmull, creative head John Lasseter, and technology lead Greg Brandeau, to lead Disney's animation arm in Los Angeles (while still leading Pixar). Catmull and Lasseter were given the alternative of closing the Disney operation, but they chose to keep it alive because of its iconic stature in the history of animation.[28]

Given that the only change made to Disney animation was the addition of these three managers and the management practices they brought with them, the turnaround at Disney was stunning. No other changes were made—no work was shared, no people transferred, and no teams from one division reported to the other. The difference at the box office was immediate and powerful (figure 8-2). The three movies Disney produced before the Pixar acquisition—*Meet the Robinsons* (2007), *Bolt* (2008), and *The Princess and the Frog* (2009)—averaged revenues of $249 million each.[29] The first three movies that Disney delivered after Catmull, Lasseter, and Brandeau's new management had a chance to take effect were *Tangled* (2010), *Wreck-It Ralph* (2012), and *Frozen* (2013), which on average earned over three times the earlier group's average.[30] *Tangled* (2010) was the first Disney animated feature in more than fifteen years to gross more than $500 million, and *Frozen* (2013) was the most successful animated feature in history. *Frozen* grossed almost $1.3 billion and won for Disney its first Oscar for Best Animated Feature.[31] *Big Hero 6* (2014), which followed soon after, was also a hit.

Disney's revenues rose from $33.7 billion in 2006 to $52.5 billion in 2015. Operating income rose even faster, from $6.4 billion in 2006 to $14.7 billion in 2015. Sales at Disney parks and resorts jumped from $10.8 billion to over $15 billion between 2010 and 2014, while consumer products sales rose from $2.7 billion to $4.0 billion. These results, aided as well by the success of the company's other media businesses (ESPN, ABC, and the television stations), tripled Disney's stock price from under $30 in 2010 to over $90 at the end of 2014.

Lessons Learned

The Walt Disney Company without Walt succeeded when it was led by leaders who understood what he had built—a Third Way organization with films, especially animated films, at its creative center where they provided the stories and characters that nourished everything else. And it struggled when its leaders saw the company instead as a diversified entertainment conglomerate. While live-action movies are and always have been part of the Disney appeal, Disney began as an animation studio, and even after nearly a century, the health of that department is ultimately what drives the well-being of the whole company.

In some ways, it's hard to blame those leaders—Walker in the 1970s and Eisner in the late 1990s—who didn't understand what Walt Disney had created. The Third Way presents great challenges. As the success of the total system grew, the complexities of each complement increased, the separation of the different businesses widened, and the ability to see them as an integrated whole declined. The political infighting that characterized Disney under Eisner's reign also served to drive apart the different parts of the company. The result was an atomizing of the system into a fractious group of battling business units rather than a smoothly functioning whole, and the overall performance of the company soon declined. Maintaining an integrated set of complementary innovations requires constant vigilance and care.

The success of a Third Way project depends on maintaining a strong and vibrant core, even if that core doesn't generate the most revenue. When Disney nurtured its core by producing a stream of high-quality animated feature films, its core and complementary businesses all thrived. When it neglected the core, when it tried to live on the legacy of past success, both its core and complementary businesses suffered.

If there was one person after Walt Disney and Roy O. Disney who understood and consistently argued for the power of the system and the importance of the storytelling at its core, it was Roy E. Disney, the nephew and son of the two founders. Twice when management drifted away from Walt's vision, Roy resigned his position at the company. Ultimately, he resigned from the board as well and led efforts to replace management. After an extended battle with stomach cancer, Roy died on December 16, 2009, just short of his eightieth birthday. At a ceremony to honor him, Disney CEO Bob Iger said: "It was Roy taking all those people and animation under his wing that led to all of those great films in the mid-80s and '90s, from *The Little Mermaid* to *The Lion King* to *Aladdin* to *Beauty and the Beast.* We certainly owe Roy a great debt of gratitude for all that . . . Animation really is our crown jewel."[32]

Epilogue

Are you responsible for guarding and perhaps reviving one of your organization's crown jewels? Is it your job to make an existing, possibly tired, product fresh and useful again? If so, we urge you to resist today's siren call for radical, disruptive innovation. Let someone else in your company reinvent the future of your industry. Follow instead the lead of LEGO, Disney, and others, and stay with your current product a little longer. By using the Third Way, you may find that product contains much more value that you can extract.

Your organization––like all others, large or small, for-profit or non-profit—has a set of crown jewels, important products that helped build the organization into what it is today. For each of those products, there are customers who have depended on you for years, customers who count on you still to deliver the products and services they use and enjoy. This book is a plea to you to honor those products and respect those customers. They depend on you now and count on you for the future.

Start by visiting those long-standing customers. "Date" them, rather than fighting your competitors. Watch your products in use. How do your customers get value from your products? Where are your customers' problems and frustrations? How can you broaden and deepen your relationship with them? In particular, contact customers who have left your product. Why did they leave? What was the moment when they decided to fire your product and hire another?

When you visit those customers, you may discover a more compelling promise, as well as complementary innovations that can help you deliver on that promise. If you do, we urge you to start small and experiment with the Third Way. Try it on one important segment of your customer base before rolling it out more broadly. Clarify the key product, define the promise, and begin designing the complements. Learn what works and build from there.

The Third Way requires humility, especially among the more experienced in your company. The features and characteristics that made your product great in the past may not be enough to keep it great in the future. Those who helped build the company need to learn how your customer has evolved, and understand the danger of HIPPOs (the highest-paid person's opinion; see chapter 3). And the leaders in your company need to give the Third Way team the scope of responsibility it requires to deliver a full portfolio of complementary innovations.

If along the way, you need advice and assistance, we have created a website populated with resources designed to help you on your journey. Visit www.innonavi.com/thirdway to learn techniques, read case studies, and connect with other innovators traveling the same path.

As we've tried to make clear throughout this book, we don't believe the Third Way is the only way to innovate. But if you're in charge of an important product, we urge you to try it. We hope the sequence of decisions we've described is helpful in guiding you. This approach is not for everyone. But it may be that, right now, in your part of your organization, the Third Way is the best way for you to innovate.

NOTES

Chapter 1. How Little Innovations Produce Big Results

1. George Day, "Is it Real? Can We Win? Is it Worth Doing?" *Harvard Business Review*, December 2007, 2–12.

2. Sarah Robb O'Hagan, interview with David Robertson, *Innovation Navigation* (radio show), October 21, 2014, www.innonavi.com/sarah-robb-ohagan.

3. Mike Esterl, "Gatorade Sets Its Sights on Digital Fitness," *Wall Street Journal*, March 10, 2016, www.wsj.com/articles/gatorade-sets-its-sights-on-digital-fitness-1457640150.

4. This definition is based on the definition used in Christian Terwiesch and Karl Ulrich, *Innovation Tournaments* (Boston: Harvard Business Review Press, 2009).

5. Another way we differ from some definitions of innovation is that we believe that innovations don't have to be profitable. Some complementary innovations can be expense items. Unprofitable products or services that help make some other product profitable can be a powerful way to innovate.

6. The concept of disruptive innovation first appeared in 1995 (Joseph Bower and Clayton Christensen "Disruptive Technologies: Catching the Wave," *Harvard Business Review*, January 1995) and was fully defined and explained two years later in Clayton Christensen's 1997 best seller, *The Innovator's Dilemma*. Since then, thousands of books and articles have used the term to talk about innovation, and its meaning has evolved away from its earlier, more specific definition. Christensen and his colleagues recently published a futile attempt to retake ownership of the term, (Clayton Christensen, Michael Raynor, and Rory McDonald "What Is Disruptive Innovation?" *Harvard Business Review*, December 2015). Although we respect the effort, we believe it was completely unsuccessful.

7. This section was written with the assistance of Edmund Pribitkin, MD, of Jefferson University Hospital. Norditropin and Nordicare are registered trademarks of Novo Nordisk, and Nutropin is a registered trademark of Genentech.

8. The liquid form of Norditropin, the form used in the flex pens, was approved and launched in 2000 (http://www.accessdata.fda.gov/drugsatfda_docs/nda/2000/21-148_Norditropin.cfm). Sales data is from Roche, Annual

Report 2015 (Basel, Switzerland: F. Hoffman-La Roche Ltd, March 1, 2016), 11, www.roche.com/fb15e.pdf; and Novo Nordisk, Annual Report 2015 (Bagsværd, Denmark, 2015), 7, www.novonordisk.com/content/dam/Denmark/HQ/Commons/documents/Novo-Nordisk-Annual-Report-2015.PDF.

9. Specifics about the relative ease of use of the Novo Nordisk pen can be found in http://www.medscape.com/viewarticle/843800_4.

10. CarMax SEC filing from fiscal year ending February 29, 2016, http://d1lge852tjjqow.cloudfront.net/CIK-0001170010/30e14202-391e-4f5f-b0f9-04dfaae15796.pdf.

11. Igor Ansoff, "Strategies for Diversification," *Harvard Business Review*, 1957.

12. George Day, "Is It Real? Can We Win? Is It Worth Doing? Managing Risk and Reward in an Innovation Portfolio," *Harvard Business Review*, December 2007.

13. Robert G. Cooper, *Winning at New Products: Creating Value Through Innovation*, 4th ed. (New York: Basic Books, 2011); Karl T. Ulrich and Steven D. Eppinger, *Product Design and Development*, 5th ed. (New York: McGraw-Hill Irwin, 2012).

14. W. Chan Kim and Renée Mauborgne, *Blue Ocean Strategy: How to Create Uncontested Market Space and Make Competition Irrelevant* (Boston: Harvard Business Review Press, 2005).

15. While Kim and Mauborgne might argue that this is too narrow a characterization of their work, none of their major examples—Yellow Tail wines, Cirque du Soleil, Southwest Airlines, Intuit, EFS, and so forth—depended on new technology for their success. All were applications of existing technologies in new ways to satisfy unmet customer needs.

16. Eric Ries, *The Lean Startup: How Today's Entrepreneurs Use Continuous Innovation to Create Radically Successful Businesses* (New York: Crown Business, 2011); Steve Blank and Bob Dorf, *The Startup Owner's Manual* (Pescadero, CA: K & S Ranch, 2012).

17. Tim Brown, *Change by Design* (New York: Harper Collins, 2009).

18. Michael Porter, *Competitive Advantage* (New York: Free Press, 1985), especially 33–118.

19. For full-spectrum innovation, see George Day, *Innovation Prowess* (Philadelphia: Wharton Digital Press, 2013). For the ten types, see Larry Keeley et al., *Ten Types of Innovation* (Hoboken, NJ: Wiley, 2013).

20. Anthony W. Ulwick, *What Customers Want* (New York: McGraw-Hill, 2005); and Rita McGrath and Ian MacMillan, *Marketbusters* (Boston: Harvard Business Review Press, 2005).

Chapter 2. LEGO and Apple Computer

1. David C. Robertson with Bill Breen, *Brick by Brick: How LEGO Rewrote the Rules of Innovation and Conquered the Global Toy Industry* (New York: Crown Business, 2013).

2. We borrow the term *gospel* from Jill Lepore, "The Disruption Machine: What the Gospel of Innovation Gets Wrong," *New Yorker*, June 23, 2014.

3. David Robertson with Bill Breen, *Brick by Brick: How LEGO Rewrote the Rules of Innovation and Conquered the Global Toy Industry* (New York: Random House Business, 2013), 68, 70.

4. Ibid, 105.

5. Jeremy Reimer, "Total share: 30 years of personal computer market share figures," *Ars Technica*, Dec 15, 2005, http://arstechnica.com/features/2005/12/total-share/10/.

6. Jason Fell, "How Steve Jobs Saved Apple," *Entrepreneur*, October 27, 2011, www.entrepreneur.com/article/220604. Other data in this section taken from Brad Moon, "5 Dead Apple Products (and Why They Failed)," *Investor Place*, January 30, 2014, http://investorplace.com/2014/01/apple-products/4/#.V0-SsJMrLq0; Matthew Fitzgerald, "The iCamera: A Look Back at Apple's First Digital Camera," *CNET*, July 28, 2009, www.cnet.com/news/the-icamera-a-look-back-at-apples-first-digital-camera; Rik Myslewski, "Reliving the Clone Wars," *Macworld*, May 23, 2008, www.macworld.com/article/1133598/macclones.html; Benj Edwards, "Steve Jobs's Seven Key Decisions," *Macworld*, September 18, 2012, www.macworld.com/article/2009941/steve-jobss-seven-key-decisions.html; and Michael Arrington, "What If Steve Jobs Hadn't Returned to Apple in 1997?" *TechCrunch*, November 26, 2009, http://techcrunch.com/2009/11/26/steve-jobs-apple-1997.

7. Edwards, "Steve Jobs's Seven Key Decisions."

8. On the dropping of various products at Apple, see the following: For the Newton PDA, see ibid. For the Pippin, see Moon, "5 Dead Apple Products." For the QuickTake, see Fitzgerald, "The iCamera." For the ending of clones, see Myslewski, "Reliving the Clone Wars."

9. Edwards, "Steve Jobs's Seven Key Decisions."

10. "Fred Anderson: There Will Never Be Another Steve Jobs," *Forbes*, October 5, 2011. www.forbes.com/sites/velocity/2011/10/05/fred-anderson-there-will-never-be-another-steve-jobs/#7563cb79ed58.

11. For the 1997 figure, see Arrington, "What If Steve Jobs Hadn't Returned."

12. All Apple financial data come from Apple Corp SEC filings, 1996–2015.

13. iTunes in 2001 and 2002 was simply an MP3 management system. Apple didn't launch the online store until April 2003.

14. Apple, "Apple Presents iPod," October 23, 2001, www.apple.com/pr/library/2001/10/23Apple-Presents-iPod.html; Brian Clark, "Picking the Right MP3 Player," *CNN Money*, January 2, 2003, http://money.cnn.com/2002/12/20/pf/saving/techguide_mp3.

15. Megan Garber, "12 Years Ago: 'Apple's iPod Spurs Mixed Reactions,'" *The Atlantic*, October 23, 2013, www.theatlantic.com/technology/archive/2013/10/12-years-ago-apples-ipod-spurs-mixed-reactions/280795.

16. For Gateway's departure from retail, see James Niccolai, "Gateway to Close All Retail Shops," *PC World*, April 1, 2004, www.pcworld.com/article/115507/article.html. For IBM's departure, see Jakki Mohr, Sanjit Sengupta, and Stanley Slater, *Marketing of High-Technology Products and Innovations* (3rd edition), (Upper Saddle River, NJ: Pearson, 2010), 326.

17. Given that LEGO's turnaround followed a similar pattern to Apple Computer's and followed Apple's moves by a few years, it's very possible that LEGO followed the Apple model.

18. "iTunes Music Store Hits Five Million Downloads," Apple, June 23, 2003, www.apple.com/pr/library/2003/06/23iTunes-Music-Store-Hits-Five-Million-Downloads.html.

19. "Apple Launches iTunes for Windows," Apple, October 16, 2003, www.apple.com/pr/library/2003/10/16Apple-Launches-iTunes-for-Windows.html.

20. Ibid.

21. Tom Mainelli, "iTunes Comes to Windows," *PCWorld*, October 16, 2003, www.pcworld.com/article/112968/article.html.

22. Eric Shiu, "Factors of Market Performance of Apple iPod: A Preliminary Desk-Based Study," *Journal of Business Case Studies* 1, no. 3 (2005): 24, cluteinstitute.com/ojs/index.php/JBCS/article/download/4924/5016.

23. "Apple Unleashes 'Tiger' Friday at 6:00 p.m.," Apple, April 28, 2005, www.apple.com/pr/library/2005/04/28Apple-Unleashes-Tiger-Friday-at-6-00-p-m.html.

24. The PowerPC chips were supplied by Freescale (formerly part of Motorola) and IBM.

25. For a detailed description of how that process works, see Robertson with Breen, *Brick by Brick*, chap. 10.

26. Benj Edwards, "The Birth of the iPod," *Macworld*, October 23, 2011.

27. Apple's 2015 sales of iPhones and iPads totaled over $178 billion, compared with $25.5 billion in Mac sales.

Chapter 3. The Four Decisions and Why They're Difficult

1. Austin Ligon, interview with author, summer 2007.

2. This analogy is from W. Chan Kim and Renée Mauborgne, *The Blue Ocean Strategy: How to Create Uncontested Market Space and Make the Competition Irrelevant* (Boston: Harvard Business Review Press, 2005).

3. This sizing, while right for a suburb of Atlanta, was far too large for most other locations. And once a store was built at this scale, it was very hard to downsize. CarMax learned from this mistake and recovered, but only because AutoNation could not execute on its version of the CarMax model.

4. Thanks to Melody Ivory for this metaphor.

5. An even more vivid, but just as appropriate metaphor, is Hugh Molotsi's comparison of mature products to movie theater urinals: "Don't Let Your Product Become a Men's Restroom," *Hugh Molotsi* (blog), December 18, 2013, http://blog.hughmolotsi.com/2013/12/dont-let-your-product-become-mens.html.

6. Hugh Molotsi, "Beware of the Hippos," *Hugh Molotsi* (blog), March 13, 2014, http://blog.hughmolotsi.com/2014/03/beware-of-hippos.html.

7. This concept is similar to Ron Adner's idea of a minimum viable ecosystem, described in his book *The Wide Lens* (New York: Penguin, 2012). We avoid using his term because there is a large difference between a centrally managed family of complementary products and an ecosystem of independent entities, some of which collaborate and some of which compete. To illustrate this, compare Apple's 2001 strategy with IBM's development of the PC twenty years earlier—the example given by the author who developed the idea of the business ecosystem (James Moore, "Predators and Prey: A New Ecology of Competition," *Harvard Business Review*, May–June 1993). When IBM began producing PCs in the 1980s, it purposely created an ecosystem around a set of PC standards and allowed a diverse set of companies to join it, including Microsoft, accessory makers, and even clone makers like Compaq. This was a true ecosystem, rather than a centrally managed federation. The outcome, of course, was that it lost control of the PC operating system to Microsoft, clone makers captured its hardware sales, and it ended up not the predator but the prey.

8. Such as Robert G. Cooper, *Winning at New Products: Creating Value Through Innovation*, 4th ed. (New York: Basic Books, 2011); and Karl T. Ulrich and Steven D. Eppinger, *Product Design and Development*, 5th ed. (New York: McGraw-Hill Irwin, 2012).

9. Two of the most influential in this category are Eric Ries, *The Lean Startup: How Today's Entrepreneurs Use Continuous Innovation to Create Radically Successful Businesses* (New York: Crown Business, 2011); and Steve Blank and Bob Dorf, *The Startup Owner's Manual* (Pescadero, CA: K&S Ranch, 2012).

Chapter 4. Decision 1: What Is Your Key Product?

1. From 72 wins and 72 losses in 2013, to 66 and 78 in 2014, to 63 and 81 in 2015 (Lehigh Valley IronPigs). Attendance in 2013 was 613,075, and, in 2015, 613,815 (Lehigh Valley IronPigs).

2. "IronPigs Draw Over 600,000 Fans, Set Record," *Ball Park Digest*, Sept 14, 2016, http://ballparkdigest.com/2016/09/14/ironpigs-draw-over-600000-fans-set-record/

3. While the Mac has undergone many changes and improvements over the years, it was and still is in essence a personal computer available in desktop and laptop versions.

4. Jim Collins, "How the Mighty Fall," *Businessweek*, May 2009, www.jimcollins.com/books/how-the-mighty-fall.html.

5. Apple January 2016 earnings report.

6. The details of how blackjack is played and how doubling down works are readily available elsewhere. For those unfamiliar with the game, the key takeaway from this example is that doubling down—if done correctly—both increases your bet and increases your likelihood of winning.

Chapter 5. Decision 2: What Is Your Business Promise?

1. GoPro began selling cameras in 2004, first using film, then digital technology to capture still images. The company didn't release its first video camera until 2006.

2. GoPro 2015 Annual Report.

3. Parts of this section are based on a paper by Wharton EMBA student Jenna Stento.

4. Unless otherwise noted, all metrics and descriptions in the GoPro story reflect the state of affairs as we write this in 2016.

5. GoPro 10K, Fiscal Year ending December 31, 2015, http://investor.gopro.com/financials-and-filings/sec-filings/sec-filings-details/default.aspx?FilingId=11223697.

6. You Tube channel websites, extracted November 2015, www.youtube.com/user/ActionCamfromSony and www.youtube.com/user/GoProCamera.

7. Unfortunately, we know this from personal experience, having bought two of these competitive cameras.

8. For GoPro numbers, see GoPro 10K, Fiscal Year ending December 31, 2015. Sony attributes declines to "unit sales of digital cameras and video cameras reflecting a contraction of these markets." Sony Financial Reports, Fiscal Year Ending March 31, 2015, *Morningstar*, March 31, 2015, http://quote.morningstar.com/stock-filing/annual-report/2015/3/31/t.aspx?t=:sne&ft=20-f&d=cd69219a7f cd78684d351b3c67fb4c45; and Sony, "Consolidated Financial Results for the Fiscal Year Ended March 31, 2015," www.sony.net/SonyInfo/IR/library/fr/14q4_sony.pdf. Sony reorganized itself in 2011, so financial data for 2012 and after cannot be compared to data before 2012.

9. Investor's Business, "GoPro Dominates Do-It-Yourself Action Video Industry," NASDAQ, November 14, 2014, www.nasdaq.com/article/gopro-dominates-do-it-yourself-action-video-industry-cm414301#/ixzz3qp7KQuKY.

10. Marc Graser, "GoPro Sees Future as Content Company," *Variety*, May 2014, www.nasdaq.com/article/gopro-dominates-do-it-yourself-action-video-industry-cm414301#/ixzz3qp7KQuKY.

11. Some readers may notice that our concept of a promise is similar to the concept of a value proposition. We believe that a value proposition is part of a promise, but not all. A promise is a value proposition coupled with a commitment to deliver that value.

12. Material in this section was taken from Jim Stengel, *Grow: How Ideals Power Growth and Profit at the World's Greatest Companies* (New York: Crown Business, 2011), 166–199.

13. This human-centered design is the heart of an approach to innovation that's come to be called *design thinking*. See Tim Brown, "Design Thinking," *Harvard Business Review*, June 2008. But there are literally hundreds of books available to help the innovation team understand customer needs. Our favorites are Christian Madsbjerg and Mikkel B. Rasmussen, *The Moment of Clarity: Using the Human Sciences to Solve Your Toughest Business Problems* (Boston: Harvard Business Review Press, 2014); Warren Berger, *A More Beautiful Question: The Power of Inquiry to Spark Breakthrough Ideas* (New York: Bloomsbury USA, 2014); and Anthony W. Ulwick, *What Customers Want: Using Outcome-Driven Innovation to Create Breakthrough Products and Services* (New York: McGraw-Hill Education, 2005).

14. Bob Wells, senior vice president of communications, Sherwin-Williams, quoted in Doug Sundheim, *Taking Smart Risks* (New York: McGraw-Hill Education, 2013).

15. We first heard this term from Bob Sutton at Stanford University.

16. The T-shaped structure of figure 5-3 is deliberate and is meant to draw a parallel to the concept of the T-shaped employee. Many companies, including the design firm IDEO, encourage their employees involved in innovation to develop a T-shaped profile. The T-shape represents a broad understanding of all the phases of their work, and a deep expertise in one or a few areas. We use the same visual metaphor to emphasize the need for innovators to not only develop a complete understanding of their customers' activity chains but also to analyze and understand their own company's value chain.

17. Theodore Levitt, "Marketing Myopia," *Harvard Business Review*, July–August, 1960.

18. For more on the jobs-to-be-done approach, see Dorothy Leonard, "The Limitations of Listening," *Harvard Business Review*, January 2002; and

Anthony W. Ulwick, "Turn Customer Input into Innovation," *Harvard Business Review*, January 2002.

19. To learn more about this type of research, we recommend the work of Tony Ulwick and his colleagues. His books and articles are full of detailed advice and recommendations that can help you through this phase of the process. See Ulwick, *What Customers Want*; and Bettencourt and Ulwick, "The Customer-Centered Innovation Map," *Harvard Business Review*, May 2008.

20. The author recently worked with a company that makes automotive entertainment and communication systems. As the systems have become more sophisticated and complex, the controls have become more difficult to use. Following the customer to understand what customers try to do with their cars while they're commuting to work is a large and complex research task, but one that yielded some deep insights into the design of the systems and their controls. This is very different from the research that the CarMax team did.

21. See Ian MacMillan and Rita McGrath, "Discovering New Points of Differentiation," *Harvard Business Review*, July–August 1997; and Ian MacMillan and Rita McGrath, *Marketbusters* (Boston: Harvard Business School Press, 2005).

22. This is essentially the value chain described by Michael Porter, *Competitive Advantage* (New York: Free Press, 1985), especially 33–118.

23. Ibid.; and Larry Keeley et al., *Ten Types of Innovation* (Hoboken, NJ: Wiley, 2013).

24. This section was coauthored with Pauline Francis, a former student in Wharton's Executive MBA program.

25. Christopher Palmeri, "Victoria's Secret Is Sexy Again," *Bloomberg Businessweek*, December 4, 2006, www.bloomberg.com/bw/stories/2006-12-04/victorias-secret-is-sexy-againbusinessweek-business-news-stock-market-and-financial-advice.

26. Michael J. Silverstein, Neil Fiske, and John Butman, *Trading Up: Why Consumers Want New Luxury Goods—and How Companies Create Them* (New York: Penguin, 2008).

27. Anna Tomasino, *Discovering Popular Culture* (New York: Pearson Longman, 2007).

28. Melanie Wells, "Cosmetic Improvement," *Forbes*, November 13, 2000, www.forbes.com/global/2000/1113/0323030a.html.

29. Ryan Faughnder, "Lingerie Retailer Frederick's of Hollywood Seeks Suitors," *Los Angeles Times*, June 2, 2012, http://articles.latimes.com/2012/jun/02/business/la-fi-0602-fredericks-hollywood-20120602.

30. Ashley Lutz, "Why Victoria's Secret Took Over the World While Frederick's of Hollywood Failed," *Business Insider*, September 10, 2012, www.businessinsider.com/victorias-secret-fredericks-of-hollywood-2012-9.

31. Samantha Masunaga, "Frederick's of Hollywood Closes All Stores, Strips Down to Web," *Los Angeles Times*, April 16, 2015, www.latimes.com/business/la-fi-fredericks-stores-closing-20150416-story.html.

Chapter 6. Decision 3: How Will You Innovate?

1. It's not a coincidence that Newell and Harrington were from the operating system group within Microsoft. What Valve has become is essentially an operating system for gamers and their games.

2. For a list of awards and ratings, see USAA, "Awards and Rankings," accessed October 18, 2016, www.usaa.com/inet/wc/about_usaa_corporate_overview_awards_and_rankings.

3. See Jeanne W. Ross and Cynthia M. Beath, "USAA: Organizing for Innovation and Superior Customer Service," working paper 382, Center for Information Systems Research, Sloan School of Management, Massachusetts Institute of Technology, December 2010, dspace.mit.edu/bitstream/handle/1721.1/68555/USAA%20RossBeath2.pdf.

4. There is no definitive study on what percentage of new products fail. Many studies report numbers as high as 75 to 95 percent, and others report failure rates in the 30–40 percent range. The difference between these rates is often one of definition. In some studies, a product is considered a failure if it doesn't reach $7.5 million in sales during its first year (Joan Schneider and Julie Hall, "Why Most Product Launches Fail," *Harvard Business Review*, April 2011). In others, a product is only considered a failure if it is pulled off the market after repeated attempts to improve it. For a review of the literature, see George Castellion and Stephen Markham, "Myths About New Product Failure Rates," *Product Innovation & Management* 30 (2013): 976–979.

5. Laura J. Kornish and Karl T. Ulrich, "The Importance of the Raw Idea in Innovation: Testing the Sow's Ear Hypothesis," working paper, Social Science Research Network, October 2012.

6. There are many good books and articles about how to construct pretotypes. Two of our favorites are Alberto Savoia, *Pretotype It!*, www.pretotyping.org; and Jeremy Clark, *Pretotyping @ Work*, PretotypeLabs, 2012, https://docs.google.com/file/d/0B0QztbuDlKs_bHdnQ2h5dnNvcE0/edit.

7. See interview with Warby Parker's founders: Eric Johnson, "How Warby Parker Learned to Love Good Old-Fashioned Retail Stores," *Recode*, April 11, 2016, www.recode.net/2016/4/11/11586024/warby-parker-dave-gilboa-neil-blumenthal-podcast.

8. We especially like Clark, *Pretotyping @ Work*.

9. Hal Gregersen, interview with the author, *Innovation Navigation* radio program, March 2014.

Chapter 7. Decision 4: How Will You Deliver Your Innovations?

1. The source for this case study is a pair of interviews: Donal Ballance and Darren Fagan, interview with author, *Innovation Navigation* radio program, October 13, 2015. Ballance and Fagan are two executives involved in the Irish Pub Concept.

2. Not the worst job in the world.

3. Guinness, now part of the British company Diageo PLC, now contracts through third-party consultants to deliver this service.

4. Based on a private conversation with a Sony executive in 2015.

5. Sony did develop some of these items—mounts and software, for example—but they were not as extensive, were not well marketed, and were not easily available in retail outlets.

6. This matrix is not the LEGO Group's actual matrix. The actual matrix is confidential. The one presented here is the author's estimate of what that matrix might have looked like.

7. See The Ellen Show, "Kevin the Cashier at the LEGO Store," video uploaded February 4, 2014, www.youtube.com/watch?v=oHhATQDcQ2k. We have no inside information, but wouldn't be surprised if LEGO had to pay the *Ellen DeGeneres Show* to make this happen.

8. LEGO describes Rebrick (www.rebrick.com) as a "social sharing platform" and hosts a series of contests on the site.

9. The contest winner (LEGO, "Gorgy Wants a Horse: *The LEGO Movie* Competition 1st Place Winner," video uploaded May 13, 2013, www.youtube.com/watch?v=AHMUx1ODdn8) can be seen briefly in the movie about twenty-two minutes before the end.

10. The exception to this is the characters that LEGO licensed from other companies. Batman, for example, plays a prominent role in the movie, so any revenues from licensing the LEGO Batman character would certainly be shared.

11. Dan Lin, interview with author, *Innovation Navigation* radio program, March 4, 2014.

Chapter 8. Lessons from an American Icon

1. As we discussed in chapter 4, sometimes the key product represents the core of who you are as a company, while at other times key products will be less central to your business. In Disney's case, we believe that one of its key products—animated feature films—was and still largely is the core of the Disney-branded part of the company. Another key product—the ESPN network and all its complementary products and services—represents another Third Way example, a story we'll leave for another time.

2. Neal Gabler, *Walt Disney: The Triumph of the American Imagination* (New York: Vintage, 2007; reprint edition), 150.

3. For a fascinating discussion of the animation process, see this article from the January 1938 issue of *Popular Science Monthly* about the making of *Snow White and the Seven Dwarfs*: http://blog.modernmechanix.com/the-making-of-snow-white-and-the-seven-dwarfs/4/#mmGal.

4. Gabler, *Walt Disney*, 123.

5. Ibid., 158.

6. Ibid., 139–141.

7. The sales of the Mickey and Minnie handcar were so large they pulled Lionel—the toy train company—out of bankruptcy.

8. Gabler, *Walt Disney*, 197–198.

9. Ibid., 277.

10. Ibid., 533.

11. Ibid., 502.

12. Ibid., 507.

13. Ibid., xiv.

14. From William Silvester, *Saving Disney: The Roy E. Disney Story* (Theme Park Press, 2015), ch. 6.

15. In spite of the mantra, Epcot, as it finally opened, was a far cry from the domed community that Walt Disney had first envisioned.

16. *Wikipedia*, s.v. "List of Walt Disney Pictures films," last modified November 7, 2016, http://en.wikipedia.org/wiki/List_of_Walt_Disney_Pictures_films.

17. James Stewart, *Disney War* (New York: Simon & Schuster, 2005), 44.

18. Ibid.

19. Caitlin Roper, "*Big Hero 6* Proves It: Pixar's Gurus Have Brought the Magic Back to Disney Animation," *Wired Magazine*, October 21, 2014.

20. Silvester, *Saving Disney*, ch. 6.

21. Walt Disney Company annual reports.

22. Roper, "*Big Hero 6* Proves It."

23. Ibid.

24. For a nice description of this relationship and how it ended, see Bruce Orwall and Nick Wingfield, "The End: Pixar Breaks Up with Distribution Partner Disney," *Wall Street Journal*, January 30, 2004.

25. ESPN, after its acquisition by Disney, became another excellent example of the Third Way.

26. Orwall and Wingfield, "The End."

27. Roper, "*Big Hero 6* Proves It."

28. Ibid.

29. Although these three movies were released after the acquisition of Pixar, they were far along in their development and the new management was unable to influence their direction to any great extent.

30. We're skipping a few films in between these two sets of movies. It's not clear who should get the credit (or, more accurately, the blame) for *Mars Needs Moms* (2011), *The Winnie the Pooh Movie* (2011), or *Frankenweenie* (2012).

31. The first Academy Award for Best Animated Feature was awarded in 2001 to *Shrek*. Of the twelve Oscars awarded in the category between 2001 and 2012, Pixar films won seven. Disney won none.

32. Silvester, *Saving Disney*, epilogue.

INDEX

ACKNOWLEDGMENTS

This book has benefited from the thoughtful critiques of many smart people over the past few years. In particular, my students in the Wharton EMBA program helped shape and refine the ideas in this book. Many thanks to Pauline Francis, Melody Ivory, Jon Jones, Sonny Kwok, Edgar Ore, Khoabane Phoofolo, Gabriella Sherman, Timaj Sukker, Whitney Moore, Viraj Narayanan, David Roberts, Akihiro Sato, Jenna Harrington, Chris Hensman, Sumeet Khanna, Anand Prabhakar, and especially Jory Lamb and Ed Pribitkin from the Wharton EMBA class of 2015 for their thoughtful additions and critiques. Prateek Bhargahva, McKenzie Brooker, James Cleveland, Amber Czonka, Stuart Gold, Kathy Huang, Nicole Metcho, Matt Panas, and Karen Stablein from the Wharton WEMBA class of 2016 also provided helpful advice to a later draft of the book. My thanks also to four anonymous reviewers who provided helpful feedback on an earlier draft, and to Ben Droz and Landon Echols for their research along the way.

I am also indebted to the many executives and academics featured in this book. Sarah Robb O'Hagan, Donal Ballance, Mick Simonelli, Rahul Kapoor and Karl Ulrich provided helpful feedback, although any errors in the book are mine alone.

And a special thank you to Katherine Jansen and Mark Adamczyk—two great friends who, despite my asking them to read a poorly written early version of the book, still remain friends. There's no better way to get harsh critiques of ideas and writing than on a bike, with smart friends, in beautiful weather. To Mark and KJ: thank you.

I'm also indebted to Jeff Kehoe, Julie Devoll, Kenzie Travers, Dave Lievens, and Patricia Boyd at HBR Press for everything they've added to the book. Many thanks also to Carol Franco, my friend and agent, for all her ideas, advice, and support over the past few years.

This book would not have been possible without the wonderful writing and thoughtful feedback that Kent Lineback provided. Kent is not only a great writer, but also a master of the craft of book writing. Organizing the ideas in a book, supporting them with data and case studies, and making it all accessible and readable is a skill that Kent is superb at both doing and explaining. Throughout the process, his careful organization, intelligent critiques, and flawless BS detection turned a rough jumble of ideas into the book you hold in your hands.

Writing a book is neither easy nor quick; it can take its psychological toll not only on the author but on his family as well. A special thanks to the love of my life, my wife Anne, for her understanding, support, and empathy over the past few years. You made writing this book much easier, Anne. Thank you.

ABOUT THE AUTHORS

DAVID ROBERTSON is a Professor of Practice at the Wharton School at the University of Pennsylvania, where he teaches innovation and product design to Wharton's executives, MBAs, and undergraduates. Robertson is also the host of the weekly radio show and podcast *Innovation Navigation*, in which he interviews leaders from around the world about the management of innovation. He is the author of *Brick by Brick*, an inside look at LEGO's near death and spectacular rebirth. Robertson is a frequent speaker at conferences and company events and serves as a consultant to companies on the best ways to structure and lead innovation. Prior to Wharton, Robertson was the LEGO Professor of Innovation at the Swiss business school IMD, an executive at four enterprise software companies, and an associate partner at McKinsey & Company. Robertson lives with his wife and two children in Chestnut Hill, a suburb of Philadelphia. For more information, see www.robertsoninnovation.com.

Before becoming a bestselling business book author, collaborator, and coach, **KENT LINEBACK** served more than twenty-five years as a manager and executive in organizations of all kinds—public, private, not-for-profit, and government. In those leadership roles, he piloted the rapid growth of a highly successful internal startup, oversaw ongoing operations, and guided a midsize public company through strategic change. As an author for the past twenty years, he has written or collaborated on seventeen books and coauthored three *Harvard Business*

Review articles, one of which won the Warren Bennis Prize for the best article on leadership in HBR in 2014. He has coauthored two books with Harvard Business School Professor Linda Hill: *Being the Boss: The 3 Imperatives for Becoming a Great Leader* and *Collective Genius: The Art and Practice of Leading Innovation*, published by Harvard Business Review Press in 2011 and 2014, respectively.